A C S   S Y M P O S I U M   S E R I E S **575**

Polymers from Agricultural Coproducts

Marshall L. Fishman, EDITOR
Agricultural Research Service,
U.S. Department of Agriculture

Robert B. Friedman, EDITOR
American Maize-Products Company

Samuel J. Huang, EDITOR
University of Connecticut

Developed from a symposium sponsored
by the Division of Agricultural and Food Chemistry
at the 206th National Meeting
of the American Chemical Society,
Chicago, Illinois,
August 22–27, 1993

American Chemical Society, Washington, DC 1994

Seplac
Chem

Library of Congress Cataloging-in-Publication Data

Polymers from agricultural coproducts / Marshall L. Fishman, editor,
Robert B. Friedman, editor, Samuel J. Huang, editor.

　　p.　　cm.—(ACS symposium series, ISSN 0097–6156; 575)

"Developed from a symposium sponsored by the Division of
Agricultural and Food Chemistry at the 206th National Meeting of the
American Chemical Society, Chicago, Illinois, August 22-27, 1993."

Includes bibliographical references and indexes.

ISBN 0–8412–3041–2

1. Polymers—Congresses.　2. Biopolymers—Congresses.　3. Biomass
chemicals—Congresses.

I. Fishman, Marshall L., 1937-　. II. Friedman, Robert B., 1938-　.
III.　Huang, Samuel J., 1937-　. IV. American Chemical Society.
Division of Agricultural and Food Chemistry. V. Series.

TP1081.P66　1994
668.9—dc20　　　　　　　　　　　　　　　　　　94–31143
　　　　　　　　　　　　　　　　　　　　　　　　　　　CIP

The paper used in this publication meets the minimum requirements of American National
Standard for Information Sciences—Permanence of Paper for Printed Library Materials, ANSI
Z39.48–1984. ∞

PRINTED IN THE UNITED STATES OF AMERICA

TP1081
P66
1993
CHEM

Foreword

THE ACS SYMPOSIUM SERIES was first published in 1974 to provide a mechanism for publishing symposia quickly in book form. The purpose of this series is to publish comprehensive books developed from symposia, which are usually "snapshots in time" of the current research being done on a topic, plus some review material on the topic. For this reason, it is necessary that the papers be published as quickly as possible.

Before a symposium-based book is put under contract, the proposed table of contents is reviewed for appropriateness to the topic and for comprehensiveness of the collection. Some papers are excluded at this point, and others are added to round out the scope of the volume. In addition, a draft of each paper is peer-reviewed prior to final acceptance or rejection. This anonymous review process is supervised by the organizer(s) of the symposium, who become the editor(s) of the book. The authors then revise their papers according to the recommendations of both the reviewers and the editors, prepare camera-ready copy, and submit the final papers to the editors, who check that all necessary revisions have been made.

As a rule, only original research papers and original review papers are included in the volumes. Verbatim reproductions of previously published papers are not accepted.

M. Joan Comstock
Series Editor

Contents

Preface

For THE LAST SEVERAL YEARS there has been a growing consensus among the industrialized nations of the world of the need for research to convert underutilized agricultural coproducts and surplus crops into new uses that add value to agricultural commodities. It is projected that new use research will provide new markets for low-valued agricultural commodities. Also the research could enhance rural development because new industries may seek to locate close to the supply of newly developed raw materials. Furthermore, new-use research that reduces crop surpluses will reduce land set aside by the governments, thereby releasing tax dollars for other programs. Another expected benefit will be to produce raw materials from sustainable and renewable resources rather than to deplete finite petroleum resources. These resources also are needed as sources of energy. Moreover, many of the newly developed materials will replace materials that degrade slowly when composted. Many of the biopolymers derived from agricultural commodities often biodegrade rapidly when composted.

This book as well as the symposium upon which it was based was assembled to gather together leading scientists and technologists working in the field of polymers derived from agricultural materials. This field will be advanced through interchange of sound research ideas from a group of researchers with diverse interests and training who might not otherwise interact.

Much of the impetus for this research occurred when it became apparent to governments worldwide that greater support for research in the field of agriculturally derived polymers could stabilize farming communities that were destabilized by worldwide competition by promoting greater efficiency. Also contributing to the revival of interest in polymers from agriculture is the desire to produce "more environmentally friendly plastics," to conserve nonrenewable resources, and to reduce trade deficits by reducing the importation of foreign oil.

The diversity of polymers from agriculture is reflected in the diversity of the book's contents. The first section contains two overview chapters, which deal with rationales and economic prospects for producing various polymers from agricultural materials, as well as the status of the technology to produce degradable polymers and applications. The last three sections of the book detail the characterization, synthesis, modification, and isolation of specific groups of polymers and polymeric materials, including starches, starch blends, composites, other polysaccharides, oligosaccharides, monosaccharides, polyamides, proteins, polyesters, and rubbers. The unifying theme of this book is the production of these polymers from sustainable and renewable agricultural resources.

The editors offer their appreciation and gratitude to participants in the symposium and to authors and reviewers of chapters, whose hard work made possible the symposium and this book.

MARSHALL L. FISHMAN
U.S. Department of Agriculture
Agricultural Research Service
Eastern Regional Research Center
600 East Mermaid Lane
Philadelphia, PA 19118

ROBERT B. FRIEDMAN
American Maize-Products Company
1100 Indianapolis Boulevard
Hammond, IN 48320

SAMUEL J. HUANG
University of Connecticut
Materials Science Institute
97 North Eagleville Road
U-136
Storrs, CT 06269

July 29, 1994

OVERVIEWS

Chapter 1

Polymeric Materials from Agricultural Feedstocks

Ramani Narayan

Michigan Biotechnology Institute, 3900 Collins Road, Lansing, MI 48910

Agricultural feedstocks should be used for the production of materials, especially plastics, and chemicals because of the abundant availability of agricultural feedstocks, the value it would add to the U.S. economy, and the reduction in U.S. trade deficit that could be achieved. However, the use of agricultural feedstocks for producing plastics, coatings and composites is negligible. New environmental regulations, societal concerns, and a growing environmental awareness throughout the world are triggering a paradigm shift towards producing plastics and other materials from inherently biodegradable, and annually renewable agricultural feedstocks. Potential plastic markets for polymeric materials based on agricultural feedstocks and the rationale for developing such materials are discussed. Technologies for using starches, cellulosics, other polysaccharides, seed oils, proteins and natural fibers in plastics related applications are reviewed.

There is an abundance of natural, renewable biomass resources as illustrated by the fact that the primary production of biomass estimated in energy equivalents is 6.9 x 10^{17} kcal/year (1). Mankind utilizes only 7% of this amount, i.e. 4.7 x 10^{16} kcal/year. In terms of mass units the net photosynthetic productivity of the biosphere is estimated to be 155 billion tons/year (2) or over 30 tons per capita and this is the case under the current conditions of non-intensive cultivation of biomass. Forests and crop lands contribute 42 and 6%, respectively, of that 155 billion tons/year. The world's plant biomass is about 2 x 10^{12} tons and the renewable resources amount to about 10^{11} tons/year of carbon of which starch provided by grains exceeds 10^9 tons (half which comes from wheat and rice) and sucrose accounts for about 10^8 tons. Another estimate of the net productivity of the dry biomass gives 172 billion tons/year of which 117.5 and 55 billion tons/year are obtained from terrestrial and aquatic sources, respectively (3).

0097–6156/94/0575–0002$09.26/0

Forests cover one third of the land in the 48 contiguous states (759 MM acres) and commercial forests make up about 500 MM acres. Fortunately, we are growing trees faster than they are being consumed, although sometimes the quality of the harvested trees is superior to those being planted. Agriculture uses about 360 MM acres of the 48 contiguous states, and this acreage does not include idle crop lands and pastures. Again, these figures clearly illustrate the potential for biomass utilization in the U.S. (3).

However, Federal farm programs idle 15 to 20% of U.S. cropland. Today, much of this land is tied up in the Conservation Reserve Program (CRP) and to build back our supplies from the effects of the year's poor harvest. But as supplies are restored and CRP ends, the long-term capacity dilemma will be with us again.

It is estimated that U.S. agriculture accounts directly and indirectly for about 20% of the GNP by contributing $ 750 billion to the economy through the production of foods and fiber, the manufacture of farm equipment, the transportation of agricultural products, etc.   It is also interesting that while agricultural products contribute to our economy with $ 40 billion of exports, and each billion of export dollars creates 31,600 jobs (1982 figures), foreign oil imports drain our economy and make up 23% of the U.S. trade deficit (U.S. Department of Commerce 1987 estimate)

Given these scenarios of abundance of biomass feedstocks, the value added to the U.S. economy, and reduction in U.S. trade deficit, it seems logical to pursue the use of agricultural feedstocks for production of materials, chemicals and fuels (4).

Biomass derived materials are being produced at substantial levels. For example, paper and paperboard production from forest products was around 139 billion lb. in 1988 (5), and biomass derived textiles production around 2.4 billion lb. (6). About 3.5 billion pounds of starch from corn is used in paper and paperboard applications, primarily as adhesives (7). However, biomass use in production of plastics, coatings, resins and composites is negligible. These areas are dominated by synthetics derived from oil and represent the industrial materials of today.

This chapter reviews the production of polymeric materials from agricultural feedstocks for applications in plastics, coatings, and composites. Subsequent chapters in the book showcase emerging polymeric materials technologies based on agricultural feedstocks. Figure 1 shows the various agricultural feedstocks available for production of polymeric materials.

Drivers for Production of Polymeric Materials Based on Agricultural Feedstocks

New environmental regulations, societal concerns, and a growing environmental awareness throughout the world have triggered a paradigm shift in industry to develop products and processes compatible with the environment. This paradigm shift has two basic drivers:

- Resource conservation/depletion -- utilization of annually renewable resources as opposed to petroleum feedstocks and the potential environmental and economic benefits that go with it.
- Environmental Concerns. -- products and processes that are compatible with the environment. Compatibility with the environment ties into the issue of waste management, that is, disposing our waste in an environmentally and ecologically sound manner. This brings up questions of recyclability and biodegradability of materials and products.

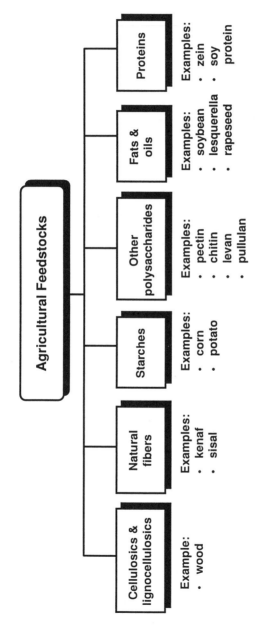

Figure 1. Agricultural feedstocks available for the production of polymeric materials.

Key international companies and industrial organizations meeting in Rotterdam recently endorsed a set of principles and a charter that will commit them to environmental protection into the 21st century (8). Some of the key principles of the charter are:

- Develop and operate facilities and undertake activities with energy efficiency, **sustainable use of renewable resources** and waste generation in mind.
- Conduct or support research on the impact and ways to minimize the impacts of raw materials, products or processes, emissions and wastes.
- Modify the manufacture, marketing, or use of products and services so as to prevent serious or irreversible environmental damage. Develop and provide products and services that do not harm the environment.
- Contribute to the transfer of environmentally sound technology and management methods.

The International Standards Organization (ISO) has formed a technical committee (ISO/TC 207) to address standardization in the field of environmental management and brings to the forefront the need for industry to address issues relating to how their products and processes impact the environment. It is anticipated that these standards will impact the industry similar to the impact of the ISO 9000 quality assurance standards.

Polymer materials derived from agricultural feedstocks can play a major role under this heightened environmental climate. Clearly, the processes, products and technologies adopted and developed utilizing renewable resources will have to be compatible with the environment. Furthermore, the wastes generated should be recycled or transformed into environmentally benign products.

The timing is right for polymer materials (plastics) and products designed and engineered from agricultural feedstocks to enter into specific markets currently occupied by petroleum based feedstocks. However, displacing a high-sales, low-cost material like plastics, that are produced by a process that operates profitably in an vertically integrated industry, is difficult. The problem is compounded by the fact that the capital for these plants has been depreciated already and they continue to operate profitably.

"Cradle to Grave" Design of Plastics

Today's plastics are designed with little consideration for their ultimate disposability or the impact of the resources (feedstocks) used in making them. This has resulted in mounting worldwide concerns over the environmental consequences of such materials when they enter the waste stream after their intended uses. Of particular concern are polymers used in single use, disposable plastic applications. Plastics are strong, light-weight, inexpensive, easily processable and energy efficient. They have excellent barrier properties, are disposable and very durable. However, it is these very attributes of strength and indestructibility that cause problems when these materials enter the waste stream. In the oceans, these light-weight and indestructible materials pose a hazard to marine life. This resulted in the Marine Plastic Pollution Research and Control Act of 1987 (Public Law 100-230) and the MARPOL Treaty. Annex V of the MARPOL Treaty prohibits "the disposal of all plastics including but not limited to synthetic ropes, synthetic fishing nets, and garbage bags". U.S. Environmental

Protection Agency (US EPA) estimates that 4,205 metric tons of plastic wastes are produced each year aboard government ships

Therefore, there is an urgent need to redesign and engineer new plastic materials that have the needed performance characteristics of plastics but, after use, can be disposed in a manner that is compatible with the environment. Thus, the twin issues of recyclability and biodegradability of polymeric materials are becoming very important . It is also important to have appropriate waste management infrastructures that utilize the biodegradability or recyclability attributes of the materials, and that these materials end up in the appropriate infrastructure. This leads us to the concept of designing and engineering new biodegradable materials -- materials that have the performance characteristics of today's materials, but undergo biodegradation along with other organic waste to soil humic materials (compost). Plowing the resultant compost into agricultural land enhances the productivity of the soil and helps sustain the viability of micro and macro flora and fauna (biological recycling of carbon).

This "cradle to grave" concept of material design, role of biodegradable polymers in waste management, and the relationship to the carbon cycle of the ecosystem have been discussed in detail by Narayan (9 -11). The production of biodegradable materials from annually renewable agricultural feedstocks for single-use disposable plastics in conjunction with composting waste management infrastructure offers an ecologically sound approach to resource conservation and material design, use, and disposal. Figure 2 shows the "cradle to grave concept" for material design from agricultural feedstocks. The concept involves integration of material redesign with appropriate waste disposal infrastructure.

Polymeric Materials (Plastics) Markets For Agricultural Polymers

The paradigm shift in material design discussed above offers new market opportunities for agricultural polymer materials. As discussed earlier, the environmental attributes of being annually renewable and biodegradable, in contrast to the current petroleum based plastics will be a major driver for entry of biodegradable plastics and other biodegradable materials based on agricultural feedstocks into the market place. However, cost and performance requirements will dictate whether or to what extent these new materials will displace current products.

Figure 3 shows the amount of thermoplastic resin sales and use by major market. As can be seen from this figure, packaging has the largest market share with 18.2 billion pounds. consumer and institutional products and adhesives/ink/coatings represent an additional 7.1 billion pounds use. Overall this represents 44.1 % of the entire plastics market. These single-use disposable plastics are not degradable and pose problems when they enter the waste stream after use. It is these plastics that have been singled out by consumers, environmentalists, legislators and regulatory agencies for attention. Thus, there is a need today to engineer single-use plastic products that have the appropriate performance properties but when disposed of in appropriate disposal infrastructures, such as composting, can biodegrade to environmentally benign products (CO_2 , water, and quality compost).

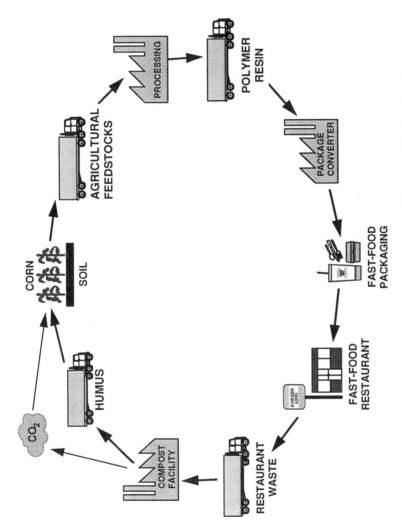

Figure 2. Cradle-to-grave concept for biodegradable materials in fast-food restaurant packaging applications.

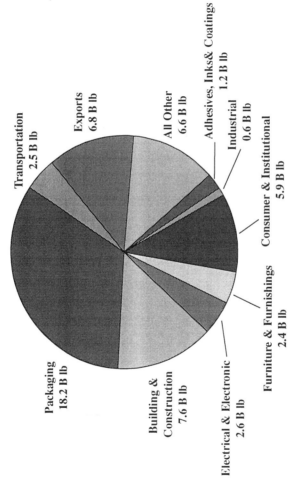

Figure 3. Thermoplastic resin sales and use by major market in 1992, billions of pounds. (Source: "Facts and Figures of the U.S. Plastics Industry, Society of the Plastics Industry, 1993)

The major target markets for biodegradable polymeric materials are:
- Single-use, disposable packaging materials
- Consumer goods -- items like cups, plates, cutlery, containers, egg cartons, combs, razor handles, toys etc.
- Disposable nonwovens (diapers, personal care and feminine hygiene products, certain medical plastics),
- Coatings for paper and film.

While biodegradable materials are not expected to completely replace all of the plastics currently used in these markets, they represent a exciting, huge business opportunity waiting to be seized. The potentially "compostable components" in the plastics and paper segments of the municipal solid waste stream representing market opportunities for biodegradable plastics is shown in Figure 4.

Packaging Resins. As discussed earlier, packaging represents, potentially, the major market for biodegradable plastics. Table I lists volume of plastic used in some disposable packaging by resin type and processing mode for 1992 and amounts to 9.3 billion pounds.

Food packaging and especially fast-food packaging is being targeted for composting because of the large volume of paper and other organic matter in the waste stream. Thus, these plastic markets would require biodegradable plastics that are compatible with the up and coming waste management infrastructure of composting. Figure 5 shows the composition of fast-food restaurant waste. It can be seen that the major component of the waste stream is readily compostable "organic waste" with a small percentage of non-biodegradable plastics. Thus, replacing the non-degradable plastics with biodegradable plastics will render this waste stream fully compostable and help convert waste to useful soil amendment. The interesting statistic shown in that figure is that 70% of customer orders are drive-thru take-out orders. As home composting grows, the demand for biodegradable plastics in these markets will increase.

Table II lists some specific, single use, disposable polystyrene market segments where the products do not lend themselves to recycling and are excellent, immediate targets for replacement by biodegradable plastics. Novon and National Starch & Chemical are already marketing starch based loose-fill packaging that is water soluble and biodegradable yet have the resilience and compressibility of polystyrene.

Non-packaging Resins. Markets for biodegradable plastics are not restricted to packaging alone. Table III shows polyethylene based non-packaging film applications amounting to 2.6 billion pounds that can potentially be captured by biodegradable plastics. In agricultural applications like mulch film, 221 MM lb. of low-density polyethylene film was used in 1991. A biodegradable agricultural mulch film would represent an energy and cost saving to the farmer because he would not have to retrieve the non-degradable film from the field. In such cases biodegradability is both a functional requirement and an environmental attribute.

The area of disposable nonwovens like diapers, personal and feminine hygiene products and certain medical plastics like face masks, gowns, gloves etc., are excellent candidates for replacement with biodegradable plastics. This is a growing market segment and these products do not lend themselves to recycling concepts.

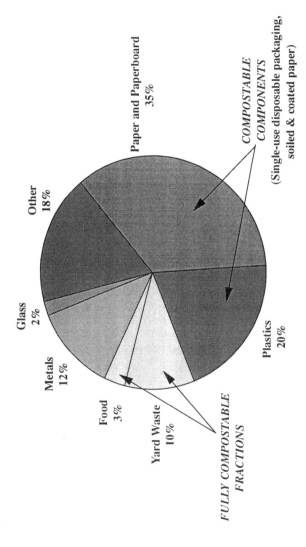

Figure 4. Compostables in municipal solid waste by volume. (Source: U.S. EPA, 1992)

Table I. Plastic Use in Disposable Applications, 1993

Application & Material type	MM lb
HD Polyethylene	
Blow molded containers	2525
Injection molded	1,099
Closures	81
Film	682
Total	**4387**
LD Polyethylene	
Blow Molded	82
Injection molded	230
Film	3740
Closures	33
Total	**4085**
Polypropylene	
Blow molded	137
Extruded	52
Injection molded	208
Thermoformed	58
Closures	426
Oriented film	512
Unoriented film	142
Total	**1535**
Polystyrene	
Blow molded	9
Molded -- solid	162
Molded -- foam	90
Thermoformed --foam	475
Thermoformed -- impact	440
Thermoformed --oriented sheet	30
Closures	201
Film	210
Total	**1617**
Polyvinylchloride	
Blow molded	195
Thermoformed	161
Closures	75
Film	240
Total	**671**
Polyethyleneterephthalate	
Blow molded	1200
Thermoformed	160
Total	**1360**
GRAND TOTAL	**9268**

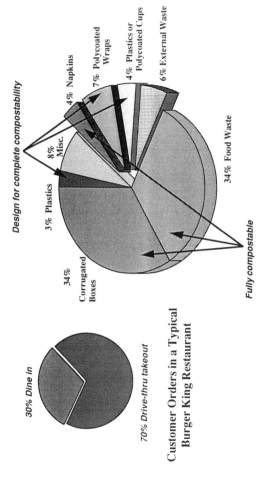

Figure 5. Composition of typical fast-food restaurant waste. (Source: The Wall Street Journal, April 17, 1991)

Table II. Specific Polystyrene Markets that are Excellent
Candidates for Biodegradable Materials

Market	MM lb
Molded articles	
Produce baskets	22
Tumblers & Glasses	80
Flatware, cutlery	90
Dishes, cups and bowls	55
Extrusion (solid) articles	
Dairy containers	142
Vending & portion cups	255
Lids	110
Plates & bowls	40
Extrusion (foam)	
Stock food trays	185
Egg cartons	55
Single-service plates	135
Hinged containers	100
Cups (non-thermoformed)	40
Expandable bead	
Packaging	101
Cups and containers	148
Loose fill	75
TOTAL	**1633**

Table III. Non-Packaging Film Markets
for Biodegradable Plastics

Market	MM lb
Agriculture	221
Diaper backing	235
Household	181
Industrial sheeting	238
Non-woven disposables	53
Trash bags	1322
Miscellanous	336
TOTAL	**2586**

Table IV lists the resins used in coatings for packaging. Coatings for paper and paperboard are excellent markets for biodegradable materials. Paper recycling and soiled paper and paperboard composting are already taking place and will grow in the years to come. A compostable/biodegradable paper coating that does not interfere in the recycling operations is needed and is being eagerly sought by manufacturers of paper products world-wide.

Nylon Resins. Nylon is a generic name for a family of long-chain polyamide engineering thermoplastics which have recurring amide groups [-CO-NH-] as an integral part of the main polymer chain. Nylons are synthesized from intermediates such as dicarboxylic acids, diamines, amino acids and lactams, and are identified by numbers denoting the number of carbon atoms in the polymer chain derived from specific constituents, those from the diamine being given first. The second number, if used, denotes the number of carbon atoms derived from a diacid. Commercial nylons are as follows: nylon 4 (polypyrrolidone)-a polymer of 2-pyrrolidone $[CH_2CH_2CH_2C(O)NH]$; nylon 6 (polycaprolactam)-made by the polycondensation of caprolactam $[CH_2\ (CH_2)_4NHCO]$; nylon 6/6-made by condensing hexamethylenediamine $[H_2N(CH_2)_6NH,]$ with adipic acid $[COOH(CH_2)_4COOH]$; nylon 6/10-made by condensing hexamethylenediamine with sebacic acid $[COOH(CH_2)_8COOH]$; nylon 6/12-made from hexamethylenediamine and a 12-carbon dibasic acid; nylon 11-produced by polycondensation of the monomer 11-amino-undecanoic acid $[NHCH_2(CH_2)_9COOH]$; nylon 12-made by the polymerization of laurolactam $[CH_2(CH_2],0CO)$or cyclododecalactam, with 11 methylene units between the linking -NH-CO- groups in the polymer chain. Typical applications for nylons are found in automotive parts, electrical/electronic uses, and packaging.

Figure 6 shows Nylon sales and use by major markets. Nylons belong to the engineering resins category and, therefore, command a premium price. As will be discussed later nylons from soy, rapeseed or lesquerella oil could potentially compete in this market and would add considerably higher value to utilization of agricultural polymer materials.

Latex Materials. Latex materials (sometimes referred to as emulsion polymers) are dispersions of the plastic polymer particles in water. Developed in the laboratory in the early 1930s, the first successful product was a synthetic rubber latex, commercialized during World War II to supplement the short supply of natural rubber latex.

A great variety and profusion of emulsions are now in commercial production. The most important of the plastic latexes are copolymers of styrene and butadiene, homopolymers and copolymers of vinyl acetate, acrylates, and vinyl chloride, as well as emulsions of polyvinyl chloride and other specialties. Other comonomers used include fumarate, maleate, and ethylene.

The non-plastic synthetic rubber latex are elastomers categorized as Styrene/Butadiene (high butadiene), Polybutadiene, Acrylonitrile/Butadiene, Chloroprene, and Butyl. There is a significant amount of inter-product competition, particularly among the plastic types, and competition with the synthetic and natural rubber lattice's that historically have been used.

The major end-uses for these plastic latex materials lie in four areas: (1) Adhesives - primarily in the packaging, construction, and wood products areas; (2) Coatings - primarily as a vehicle in water-based paints; (3) Paper - primarily as a clay coatings binder, but also used for saturating; and (4) Textiles - primarily as fabric finishes, sizes, back coats, and in fabric lamination and non-woven fiber bonding. The emulsions and compounds generally are sold or used in relatively small batches (drums, tank trucks, and tank cars), and in many applications the emulsions are tailor-made to the end-use. There are thousands of end-users, some of whom polymerize for themselves, some of whom buy compounded material.

Each product sold has some unique characteristic desired in the end-use. For example, some adhesive resins should impart water resistance. Fabric resins should be hard or soft depending on the type desired, and coatings resins should have good tint-retention properties .

As a class, these plastic emulsions have gained wide acceptance in the above-mentioned end-uses for one or more of the following reasons:

+ Water base (will meet the requirements of air pollution controls and regulations)
+ Ease of clean-up
+ Good adhesion to various substrates
+ High pigment-binding capacity
+ Abrasion resistance and flexibility
+ Uniform quality as compared with natural binders
+ Good supply picture as compared with many of the natural binders
+ Relative low cost versus binding efficiency
+ Compatibility with existing technology and processes

Figure 7 shows the styrene based latex sales and use by major market. This market is an excellent target for soy and other agricultural crop based protein fractions.

Alkyd Resins. Alkyd resins came into commercial use over 50 years ago, and even with the wide array of other polymers for coatings that have appeared in most recent years, they rank as the most important synthetic coating resin. They still constitute about 35% of all resins used in organic coatings. In those days, the plentiful supply of low-cost soybean oil and highly refined fatty acids from tall oil spurred the growth in the use of alkyd resins. The alkyds based on soybean oil gave good drying rates and good color. Alkyds are thermosetting unsaturated polyester resins produced by reacting an organic alcohol with an organic acid, dissolved in and reacted with unsaturated monomers such as styrene, diallyl phthalate, diacetone acrylamide or vinyl toluene. The typical applications are found in electrical uses, automotive parts, and as coatings. Unfortunately, the market share enjoyed by the soybean oil in the alkyd resin coatings market has remained the same since its initial introduction, while the newer synthetics have captured and expanded the alkyd resin coating markets. Over 700 MM pounds of alkyd resins are used annually.

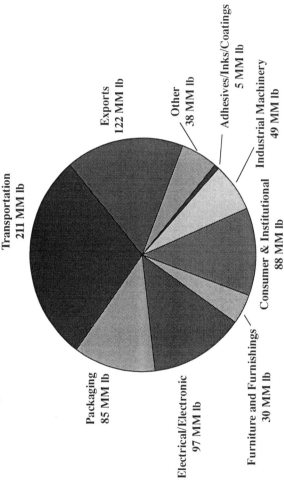

Figure 6. Nylon sales and use by major market in 1992, millions of pounds. (Source: "Facts and Figures of the U.S. Plastics Industry, Society of the Plastics Industry, 1993)

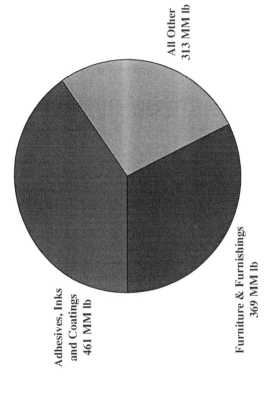

TOTAL: 11.1 billion pounds

All Other
313 MM lb

Adhesives, Inks
and Coatings
461 MM lb

Furniture & Furnishings
369 MM lb

Figure 7. Styrene-based latex sales and use by major market in 1992, millions of pounds. (Source: "Facts and Figures of the U.S. Plastics Industry, Society of the Plastics Industry, 1993).

Table IV. Potential Coatings Markets for Biodegradable Plastics

Coatings Resin	MM lb
Epoxy	36
EVA copolymer	75
Polyethylene, HD	81
Polyethylene, LD	815
Polypropylene	20
Polyvinyl acetate	48
Polyethylene terepthalate(PET)	12
Polyvinyl chloride	23
Other	96
TOTAL	**1206**

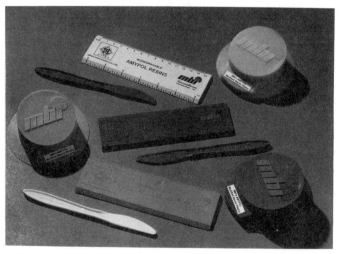

Figure 8. Biodegradable injection molded articles made from AMYPOL resin (MBI technology).

Agricultural Polymer Materials

Figure 1 shows biopolymers available from agricultural feedstocks that can potentially be converted into biodegradable plastics and environmentally friendly materials.

Polysaccharides. Starch, primarily from corn, has been the dominant agricultural biopolymer that has been targeted for conversion to biodegradable plastics. Starch is a mixture of two polysaccharides differing in structure and molecular dispersity. The two polysaccharides are:

- Amylose -- a, predominantly, linear α-(1-4) glucan and
- Amylopectin -- a highly branched α-(1-4) glucan with branch points occurring through α-(1-6) linkages.

Amylose has a molecular weight of approximately 1 million and the amylopectin molecular weight is of the order of 10 million or more. The two components, amylose and amylopectin, are present in varied ratios depending on the source.

Starch is not a thermoplastic material and degrades during processing. However, when starch is heated above the glass transition and melting temperature of its components in the presence of water and under pressure, the molecular structure of starch is disrupted and the resulting material shows thermoplastic properties. Polymer compositions containing this thermoplastic, destructrized starch have been developed for different applications and is the subject of several patents (12 ,13). Based on this technology, two companies Novon (Division of Warner Lambert) and Novamont (Italy) have introduced commercial products under the trade name Novon and Mater-Bi respectively. These starch based resins can be blown into film, injection molded, and thermoformed. The target markets for these resins were discussed in the earlier section on markets for biodegradable plastics.

Doane and coworkers present evidence (14) to show that the thermoplastic, destructurized starch is not a new or novel material but merely another term used to describe the well-known disordering of starch chains and the melting of crystallites that take place when starch is heated under pressure in the presence of limited amounts of water. In any case, the use of thermoplastic, destructurized starch, whether old or new, has been applied to the commercial production of single-use, disposable plastic films and products only recently by Novon and Novamont.

More recently, Michigan Biotechnology Institute (MBI) has signed a joint venture with Japan Corn Starch to commercialize thermoplastic, modified-starches which have water repellent properties, mechanical strength, and good processability, while being fully biodegradable in appropriate disposal systems (15). Figure 8 shows photographs of molded products based on fully biodegradable AMYPOL resins developed by MBI. National Starch and Chemical Company have developed a water soluble, fully biodegradable loose-fill (peanut) packaging material based on a low ds (degree of substitution) propoxylated high amylose starch (16). This new material is a replacement for non biodegradable expanded polystyrene (EPS) loose-fill packaging. A number of other companies, including Novon have different versions of this loose-fill peanut packaging on the market today. Even though the price of these new products are double that of EPS loose-fill, the starch based loose-fill products have already garnered 8% of the market. The total market for loose-fill packaging material in 1992 was 390 million cubic feet valued at $ 180 MM. Table V lists current biodegradable plastics producers using starch and other feedstocks.

A number of chapters in this book discuss various aspects of starch-based plastics technologies. A detailed review on producing starch-based polymeric materials has been published recently (7). Other polysaccharides such as chitin (β -1-4 linked 2-acetamido-2-deoxy-D-glucose), pullulan (maltotriose unit consisting of three α -1-4 glucosidic linkages polymerized through α -1,6 linkages on the terminal glucose residues), levan (anhydro-D-fructofuranoside units with predominantly β-2,6 glycosidic linkages and some β-2,1 branching), pectin (heteropolysaccharide, usually containing D-galacturonic acid and its methyl ester) are being studied for polymer materials applications. Coffin and Fishman discuss starch-pectin blends in this book. Pullulan films have been commercialized by a Japanese company, Hayashibara Co. Ltd.

Starch also serves as the feedstock for producing glucose that can be fermented to lactic acid. Lactic acid can be polymerized to poly (lactic acid) polymers and copolymers. The use of polylactide polymers and copolymers for biodegradable plastics is a fertile field, and considerable R&D activity is ongoing. Argonne National Laboratories has licensed their poly(lactic acid) technology to a Japanese firm, Kyowa Hakko, although commercial production is not expected for at least two to three years. The Argonne technology involves production of lactic acid by fermentation using potato waste as the feed stock. Condensation polymerization of lactic acid produces low molecular weight poly(lactic acid) which is then spliced together using coupling agents to give high molecular weight poly(lactic acid). High molecular weights are essential for good mechanical properties. The Cargill, and Ecochem technologies (Table V) involve a two-step process that converts the lactic acid to its dehydrated dimer, the lactide, followed by ring opening polymerization to high molecular weigh polylactide (PLA) polymers. Polylactide copolymers are prepared by copolymerization of other lactone monomers like glycolide, and caprolactone with the lactide monomer. Battelle has entered into a R&D joint venture with Golden Technologies Inc., Golden, Co. to explore PLA materials for commercial packaging. Polylactide polymers and copolymers are, currently, widely used in a number of biomedical applications like resorbable sutures, prosthetic devices, and as a vehicle for delivery of drugs and other bioactive agents.

Poly(hydroxybutyrate) (PHB), and poly(hydroxybutyrate-co-hydroxyvalerate) (PHBV) are novel thermoplastic polyesters that are prepared by a bacterial fermentation process using a variety of feed stocks including glucose and acetic acid . PHB is a brittle polymer. However, introducing hydroxyvalerate groups on the polymer backbone (0-30%) reduces the crystallinity, and the resultant material is much more ductile and flexible. A good balance of properties can be achieved by varying the comonomer content to yield polymers for specific applications. PHBV is currently in commercial use for blow molded shampoo bottles in the U.S., Japan, and Germany. It is also being used to make razor handles in Japan. There is a considerable body of literature in this field and the reader is referred to papers by Holmes (17)and Galvin (18) and the references cited in them. There is, also, considerable R&D activity in the field of bacterial polyesters under the general name of polyhydroxyalkanoates and is the subject of yearly International Symposia (19). Also, Koning and Lemstra discuss the prospects of bacterial poly(hydroxyalkanoates) in this book.

Cellulose is one of the most abundant constituents of biological matter. It is a polymer composed of anhydroglucose units linked by β(1-4) bonds, unlike starch where the anhydroglucose units are linked by α (1-4) bonds. Cellulose is readily biodegradable and cellophane, a regenerated cellulose film, made by the viscose process still finds many applications (20). Table V lists one of the cellophane producers. DuPont, one of the major suppliers of cellophane, is no longer making it as a result of poor economics and competition with other olefin polymers. Furthermore, the viscose process of making cellophane has negative environmental impacts. Because cellophane is inherently biodegradable and forms good films, it has much promise for film applications. However, an economical and environmentally benign process for cellophane production is needed before cellophane can capture some of the markets it lost to the non-degradable olefin polymers. Cellulose esters like cellulose acetate and mixed esters like cellulose acetate propionate (CAP) and butyrate (CAB) are thermoplastic commercial products on the market. The high degree of ester group substitution (2.4 out of a maximum of 3) renders these plastics not readily biodegradable. There is some uncertainty about the biodegradability of cellulose acetate with a ds of 2.4 and more work needs to be done in this area (21 , 22). Lower ds cellulose acetates (below 2.0) are readily biodegradable and could be potentially used for biodegradable plastics applications.

Seed Oils. American agriculture produces over 16 billion pounds of vegetable oils each year. These domestic oils are extracted from the seeds of soybean, corn, cotton, sunflower, flax, and rapeseed. Although more than 12 billion pounds of these oils are used for food products, not much usage is seen in the plastics and plasticizer fields. As discussed earlier, alkyd resins, which today constitutes about 35% of all resins used in organic coatings, was developed and grew because of the plentiful supply of low-cost soybean oil and highly refined fatty acids from tall oil. Unfortunately, the market share enjoyed by the soybean oil in the alkyd resin coatings market has remained the same since its initial introduction, while the newer synthetics have captured the alkyd resin coating markets and continue to dominate it.

Figure 9 gives the composition of crude soy oil. Both the triglycerides and the free fatty acids can be epoxidized to give epoxidized soybean oil (ESO). The transesterified fatty acid ester is currently being promoted for soy based diesel fuel applications. However, both epoxidized soybean oils can find applications in higher value-added products like composites, epoxy based thermoset materials, and as plasticizers. Fourteen companies, which includes some major names like Ferro Corporation, Union Carbide, Akzo Chemicals, Elf Atochem, Henkel, Huls-America Inc., Witco Corporation, produce epoxidized soybean oil. One hundred million pounds per year of epoxidized soybean oil finds use as a plasticizer for different plastic resins. However, this is minuscule when compared to the 1.324 billion pounds of petrochemical plasticizers that find use in the plastics area (Figure 10). Epoxidized soybean oil is the preferred plasticizer for Poly(vinyl chloride) PVC because it can

Table V. Major Biodegradable Materials Producers

Company	Base Polymer	Feedstock	Cost ($/lb)	Capacity (MM lb/yr)
Cargill, Minneapolis, MN	Polylactide (PLA)	Renewable Resources, Corn	1.00 -3.00	10 ('94 scaleup); 250 (mid-1996)
Ecochem, Wilmington, DE	Polylactide copolymers	Renewable Resources, Cheese whey, corn	< 2.00 proj'd	0.15 ('94 scaleup)
Flexel, Atlanta GA	Cellophane (Regenerated cellulose	Renewable resources	2.15	100
Zeneca (business unit of ICI)	Poly(hydroxybutyrate-co-hydroxyvalerate), PHBV	Renewable resources -- carbohydrates (glucose), organic acids	8.00 - 10.00; 4.00 proj'd	0.66, additional capacity slated for '96 is 11 - 22
Novamont, Montedison, Italy	Starch-synthetic polymer blend containing approx. 60% starch	Renewable resources + petrochemical	1.60 - 2.50	50. in Turni, Italy
Novon Products (Warner-Lambert), Morris Plains, NJ*	Thermoplastic starch polymer compounded with 5-25% additives	Renewable resources, Starch	2.00 - 3.00	100
Union Carbide, Danbury, CT	Polycaprolactone (Tone polymer)	Petrochemical	2.70	< 10
Air Products & Chemicals, Allentown, PA	Polyvinyl alcohol (PVOH) & Thermoplastic PVOH alloys (VINEX)	Petrochemical	1.0 -1.25 (PVOH); 2.50- 3.00 (VINEX)	150 - 200 (water sol. PVOH); 5 (VINEX)

Table V. *Continued*

Company	Base Polymer	Feedstock	Cost ($/lb)	Capacity (MM lb/yr)
National Starch & Chemical, Bridgewater, NJ	Low ds starch ester	Renewable resources, Starch	2.00 - 3.00	Not available
MI Biotech Inst./GRT - Japan Corn Starch Joint Venture, MI	Water repellant, thermoplastic modified starches	Renewable Resources, Starch	1.0 - 1.50	0.1(pilot scale); 150 slated for early '96
Planet Packaging Technologies, San Diego, CA	Polyethylene oxide blends (Enviroplastic)	Petrochemical	3.00	10
Showa Highpolymer Co., Ltd.	condensation polymer of glycols with aliphatic dicarboxylic acids (BIONELLE)	Petrochemical	approx. 3.00	0.2 (pilot); 7(semi-commercial, end '94)

* Warner-Lambert has recently announced the closing of its Novon Products Division.

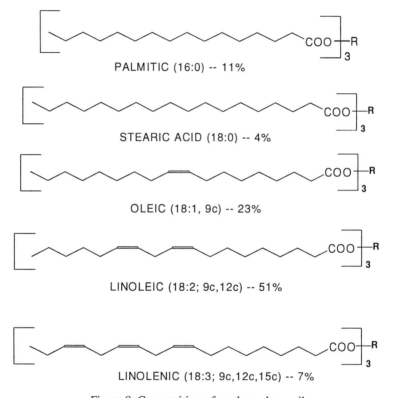

PALMITIC (16:0) -- 11%

STEARIC ACID (18:0) -- 4%

OLEIC (18:1, 9c) -- 23%

LINOLEIC (18:2; 9c,12c) -- 51%

LINOLENIC (18:3; 9c,12c,15c) -- 7%

Figure 9. Composition of crude soybean oil.

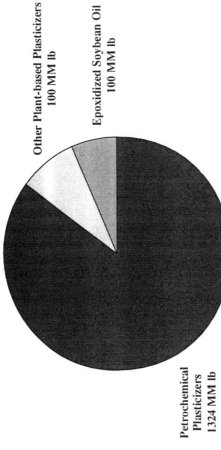

Figure 10. Plasticizer production from plant oils and petrochemical feedstocks in the U.S., millions of pounds per year. (Sources: Oil Crops of the World, 1990; Missouri Soybean Assoc., 1991; U.S. Dept. of Commerce)

confer thermal stability on the material in addition to plasticizing the material. In spite of these excellent qualities, ESO has shown very little market growth.

Nylon 9 has been successfully synthesized in the laboratory by chemically transforming soybean oil into an amino carboxylic acid (9-aminononanoic acid) followed by polymerization. Nylon 9,9 can be prepared by converting soybean oil to the C-9 dicarboxylic acid and C-9 diamine followed by polymerization. The technical feasibility has been demonstrated but no effort has been made to bring the technology to "investment grade" and attract commercial development. As discussed earlier, Nylons belong to the class of high value engineering thermoplastics ($2-3 per lb.) and command an annual market of 725 MM lb.

Another nylon polymer, Nylon-13,13 can be prepared from erucic acid which is a major constituent of industrial rapeseed oil. USDA's Cooperative State Research Service is leading an High Erucic Acid Development effort to expand commercial use of industrial rapeseed and crambe (23). Another preparation of Nylons (hydroxylated nylons) is presented by Kiely and coworkers in one of the chapters in this book.

Lesquerella is another experimental crop that is a source of oil and hydroxy fatty acids that can be used in a variety of applications, including cosmetics, waxes, nylons, plastics, coatings, and lubricants. S.F. Thames and co-workers discuss the synthesis, characterization, derivation, and application of Lesquerella and its coproducts in one of the chapters in the book.

Proteins. Soy protein fractions are currently used in adhesives and paper applications. However, recent research reports on the synthesis of protein-styrene butadiene latexes opens up the use of soyprotein in styrene based latexes (24). We, at Michigan Biotechnology Institute have developed a modified corn protein (zein) formulation with good moisture and grease barrier properties as replacement for current polyethylene and wax coatings used on paper and paperboard. The modified zein coating is biodegradable and hydrolyzable during pulping (paper recycling operations) (Michigan Biotechnology Institute, BioMaterials R&D report).

Natural fibers. In recent years there has been mounting interest in the use of natural fibers for preparing low-cost fiber reinforced thermoplastic composites for applications in the construction and automobile markets. Examples of natural fibers that can be used are wood, sisal, and kenaf. Kenaf is a new U.S. fiber crop that is grown mostly in the southern and western States (21). The advantages of using natural fibers over traditional glass fibers as reinforcement for composites are: acceptable specific strength properties, low cost, low density, high toughness, good thermal properties, reduced tool wear, low energy content and biodegradability. It has been demonstrated that wood fiber reinforced polypropylene composites has properties similar to traditional glass fiber reinforced polypropylene composites (25). Lignocellulosic-plastic composites have been recently reviewed by Rowell, Youngquist, and Narayan (4).

Conclusion

In 1769 Benjamin Franklin said " There seems to be but three ways for a nation to acquire wealth: war as the Romans did in plundering their conquered neighbors, commerce, which is generally cheating; and agriculture, the only honest way .... a kind of continual miracle ...." As society increases its understanding of the environmental implications of its industrial practice, greater attention is being given to the concept of sustainable economic systems that rely on annually renewable sources for its energy and materials. A new class of annually renewable, biodegradable materials is emerging from the transformation of agricultural feedstocks. A recent report (26) prepared by the Office of Technology Assessment, U.S. Congress highlights the potential for transforming agricultutal or marine feedstocks to this new class of renewable, biodegradable, and biocompatible materials. The report also reveals the strategic importance placed in this area by the EC countries and Japan, and the strong financial investment made by them in developing new technologies based on biopolymers.

As shown in this and other chapters of the book, the potential for utilizing agricultural feedstock is enormous and the opportunity is here. However, a strong and sustained national commitment from government, industry, and academe is needed to fulfill the expectations.

Literature Cited

1 . *Primary Productivity of the Biosphere*, Lieth, H., and Whittaker, H. R., Eds., Springer Verlag, 1975.
2 . Institute of Gas Technology, Symposium on "*Clean Fuels from Biomass and Wastes*", Orlando, Florida, 1977.
3 . Szmant, H. H., *Industrial Utilization of Renewable Resources*, Technomic Publishing Co., Lancaster, Basel, 1986.
4 . Emerging Technologies for Materials and Chemicals from Biomass, Ed., Rowell, R. M., Schultz, T. P., and Narayan, R., *ACS Symp Ser.*, 1991, 476.
5 . Cavaney, R. *Pulp Pap Int.*, 1989, 31, July, 37.
6 . Chum, H. L., and Power, A. J., *ACS Symp Ser.,* 1991, 476, 28.
7 . Doane, W. M., Swanson, C. L., and Fanta, G. F., *ACS Symp Ser.*, 1991, 476, 197.
8 . News Item, *C & E News,* April 8, 1991, pp 4.
9 . Narayan, R., *Kunststoffe,* 1989, 79 (10), 1022.
10 . Narayan, R., In *Science and Engineering of Composting*, H.A. Hoitink and H.M. Keener, Eds., pg. 339, 1993.
11 . Narayan, R., Biodegradable Plastics in *Opportunities for Innovation: Biotechnology*, National Institute of Standards and Technology (NIST, U.S. Department of Commerce) Monograph; NIST GCR-93-633, pg 135, 1993.
12 . Lay, G., Rehm, J., Stepto, R. F., Thoma, M., Sachetto, D., Lentz, J., and Silbiger, J., U.S. Pat. 5,095,054, 1992, and references cited in there.
13 . Bastioli, C., Bellotti, V., Del Giudice, L., Del Tredici, G., Lombi, R., and Rallis, R., PCT Int. Patent Appl. WO 90/10671, September 20, 1990.
14 . Shogren, R. L., Fanta, G. F., and Doane, W. M., *Starch*, 1993, 45, 276

15 . Narayan, R., *Proceedings: Third International Scientific Workshop on Biodegradable Polymers and Plastics*, Osaka, Japan, Nov. 9-11, 1993.
16 . Lacourse, N. L., and Altieri, P. A., U.S. Patent 4,863,655, 1989 & 5,035,930, 1991.
17 . P.A. Holmes, L.F. Wright, and S.H. Collins, Eur. Patent Appln. EP 69497.C., 1987.
18 . T.J. Galvin, In *"Degradable Materials: Perspectives, Issues, and Opportunities"*, S.A. Barenberg, J.L. Brash, R. Narayan, and A.E. Redpath, Eds., CRC press, FL., p.39, 1990.
19 . Proceedings of International Symposium on Bacterial Polyhydroxyalkanoates -- ISBP'92, Gottingen, Germany, June 1-5, 1992.
20 . Oswin, C. R., in *Plastics Films and Packaging*, John Wiley & Sons, New York, 1975, p 88.
21 . Narayan, R., and David, J., Bio/Environmentally Degradable Polymer Society Meeting Abstracts, June 6-8, 1994, pg 23.
22 . Buchanan, C. M., Boggs, C. M., Dorschel, D. D., Gardner, R. M., Komarek, R. J., Watterson, T. L., and White, A. W., Bio/Environmentally Degradable Polymer Society Meeting Abstracts, June 6-8, 1994, pg 12.
23 . Industrial Uses of Agricultural Materials, U.S. Department of Agriculture (USDA) report, June 1993, pg 15-17.
24 . Krinski, T. L., *ACS Symp Ser.,* 1991, 476, 299.
25 . Krishnan, M., and Narayan, R., *Mat. Res. Soc.,* Proceedings, 1992, 266, 93.
26 . U.S. Congress, Office of Technology Assessment, *Biopolymers: Making Materials Nature's Way* -- Background Paper, OTA-BP-E-102 (Washington D.C., September, 1993.

RECEIVED July 26, 1994

Chapter 2

Status of Technology and Applications of Degradable Products

Graham M. Chapman

Ecostar International, 181 Cooper Avenue, Tonawanda, NY 14150–6645

In spite of the absence of agreed standards and positive legislation, there has been a large research effort in the USA, Japan and Europe on degradable plastic materials, and the understanding of degradation mechanisms has considerably increased. The use of destructured starch, particulate starch, polyolefin degradants, polyesters (including those microbiologically produced) and petroleum based degradable polymers is described, together with their advantages, disadvantages and commercial application.

In spite of the absence of positive legislation in favor of degradable plastics, and the relevance of ASTM test methods to real environmental conditions not being established, there is a large research effort in industry, universities and other technical institutions in the USA, Europe, Japan and China. The technology is developing rapidly although extensive commercialization has not yet been achieved.

There has been legislation in Italy for degradable bags, but this lapsed in 1991. There is legislation in favor of photodegradable six-pack rings in several states in the USA and Federal legislation is proposed. The motivation for this is principally the problem of animal entrapment and injection. The recent annex to the Marpol Treaty banning the dumping of plastics at sea has also been responsible for a flurry of activity, particularly for waste bags for the US Navy.

There has been in the USA some adverse legislation against biodegradable plastics because of a perceived negative impact on recycling. This is generally illogical because most biodegradable plastics are not intended for use where recycling can be readily achieved.

0097–6156/94/0575–0029$08.00/0
© 1994 American Chemical Society

Another problem, particularly in the USA, has been the FTC and the restrictions on the claims that can be made when advertising or promoting degradable products. It is not possible to state simply "biodegradable" or "compostable" unless, in the normal disposal system, these products will biodegrade or be composted. Since many products end up in landfill, where degradation is slow, this severely restricts the claims that can be made. If composting facilities are not generally established in the area of sale of a compostable product, then this claim can only be made with qualification.

Biodegradable Plastics

For the digestion of biodegradable plastics there have been three approaches:

1. to design a product that is directly susceptible to natural biological systems,
2. to make a product which, in certain environments, can be converted to a material that is directly attacked by microbes or other natural organisms,
3. there are also combinations of the two.

It had been assumed that the synthetic and degradative pathways of natural macromolecules are the inverse of one another. A corollary of this is that only macromolecules that are produced in nature can be broken down by organisms readily available in the natural environment.

This hypothesis is challenged by recent work on the breakdown of certain hydrocarbons that are assumed could not have been produced by natural enzymatic systems. Enzyme-based reactions that involve the formation of reactive free radicals could be expected to be relatively indiscriminate and non-selective of their substrate. This has recently been confirmed by the work of Meister at the University of Detriot(1), who found that polystyrene could be degraded by white rot fungi when their traditional substrate, lignin, was present.

This illustrates the importance of recognizing the flexibility of microorganisms. It has also been shown that they have the ability to adapt to PVA of certain molecular weights(2) and it has been found that the growth of yeasts on hydrocarbons involves a high degree of metabolic specialization including the production of surface-active agents(3). This shows the difficulty of designing test methods, since certain plastics deposited in forests where white rot fungi persist may eventually degrade but the same materials in the warm compost environment, in the presence of thermophiles, may not be susceptible to degradation.

However, for certain applications, very rapid biodeterioration or biodegradation may be necessary and one starting point is to take a natural polymer, such as gelatin, cellulose or starch, and consider how these could be transformed to a practical "plastic" or packaging material.

Commercial Technologies

The main commercial technologies for biodegradable plastics are summarized in Table I.

Table I. Summary of Commercial Biodegradable Plastics Technologies.

Technology	Manufacturer/Supplier
Starch based:	
Granular starch/catalytic	Ecostar International (USA)
	Ampacet (USA)
	FCP (USA)
Destructurized/gelatinized starch with plasticizer, etc.	Novon Products (USA)
	Fluntera (CH)
	Novamont (I)
	Agritech (USA)
	American Excelsior (USA)
Starch graft copolymer	Unistar (USA)
Polyesters:	
Polylactic acid	Ecochem (USA)
	Cargill (USA)
Polycaprolactone	Union Carbide (USA)
	Cargill (USA)
Polyhydroxybutyrate/valerate (microbiologically produced)	Zeneca (GB)
Synthetic:	
Polyvinyl alcohol/ Ethylene vinyl alcohol	Air Products (USA)
	Eval/Kurary (USA,J)
Cellulose acetate	Eastman (USA)

Cellulose has been used as the basis of man-made materials for hundreds of
years. Paper is the most obvious product based on cellulose, but is not as
versatile as most synthetic plastics. Changing the properties of cellulose, e.g.
by acetylation or nitration, can produce more practical materials, but with
increasing substitution (generally above 2.4 in the case of acetylation) can result
in a material that is no longer biodegradable (4). Protein polymers, such as
gelatin, have not generally been used for packaging except for medical capsules.

Starch in its gelatinized or destructurized form has been the focus of
considerable development effort in recent years. In its gelatinized form it is
readily accessible to natural enzymes, amylases, and it is available from several
renewable plant sources. A precursor of all this current activity is rice paper, a
film based on rice which has the disadvantages of all simple starch-based
systems - water sensitivity and brittleness.

There have been two approaches to overcoming the principal inherent
problems of materials based on destructurized starch. Firstly, to add a polar
thermoplastic polymer (also fulfilling the function of a phase-transfer agent
when used in conjunction with a typical thermoplastic polymer such as a
polyolefin). Secondly, to exploit the ability of starch at certain water contents,
pressures, and temperatures to exhibit plastic qualities. These approaches are
not mutually exclusive.

The early work on the first approach was carried out by Otey and his group
at the Agricultural Research Services Northern Regional Research Center
(USDA) in Peoria. They used ethylene acrylic acid copolymer (EEA) or poly
vinyl alcohol (PA) as the polar thermoplastic polymer together with other
ingredients to produce a film (5). This technology was licensed to Agritech,
who further developed it, but it is not generally currently commercially
available, although it is being used in China.

A similar approach is that of the "Montedison Group", who also incorporated
small amounts of plasticizer to help produce a processible material. For polar
polymers they also suggest the use of ethylene vinyl acetate (EVA) and ethylene
vinyl alcohol (EVOH), (6). However, in order to achieve a practical and cost-
effective material, and to limit the product's susceptibility to moisture a greater
amount of a conventionally non-degradable polymer may be added, which can
negate the original objective.

With the second approach, originally pioneered by Professor Tomka at ETH
Zurich, injection-molded items have been made fairly readily, although the
moisture susceptibility remains a problem (7). This is the technology that
Novon (Warner-Lambert) attempted to commercialize and they have further
developed the technology, but incorporating non-readily biodegradable polymers
(7). Newer technology from Professor Tomka is proposed and utilized by
Fluntera AG, and film samples have been produced (8).

Most of the main polar polymers/phase transfer agents are summarized in Table II. Selection criteria are based on functionality and cost.

Table II. Principal ethylene copolymers used as phase transfer agents.

Name	Chemical Formulation	Typical Range of Degree of Substitution
EAA (Ethylene acrylic acid)	$-[CH_2 -CH_2 -CH_2 - CH- CH_2 - CH_2]_n -$ with branch $\overset{C}{\underset{O\,OH}{}}$	3-20%
EMAA (Ethylene methyl acrylic acid)	$-[CH_2 - CH_2 - \overset{CH_3}{\underset{}{C}} - CH_2 - CH_2]_n -$ with branch $\overset{C}{\underset{O\,OH}{}}$	3-20%
Neutralized EAA (with eg. Na^+, Zn^{++}, Li^+, Ba^+, Mg^{++} or Al^+)	$-[CH_2 - CH_2 - CH - CH_2 - CH_2]_n^-$ with branch $\overset{C}{\underset{O\,O^-\ Me^+}{}}$	3-20%
EVA (Ethylene vinyl acetate)	$-[CH_2 - CH_2 - CH - CH_2 - CH_2]_n^-$ with branch $\overset{O}{\underset{C}{}}$, $\overset{}{\underset{O\,CH_3}{}}$	2-30%
EMAC (Ethylene methylacrylate)	$-[CH_2 - CH_2 - CH - CH_2 - CH_2]_n^+$ with branch $\overset{C}{\underset{O\,OCH_3}{}}$	10-24%
EVOH (Ethylene vinyl alcohol)	$-[CH_2 - CH - CH_2]_n^-$ with branch OH	25-50%
PA (Polyvinyl alcohol)	$-[CH_2 - CH - CH_2 - CH - CH_2 - CH]_n^-$ with branches OH, OH, OH	----
EEA (Ethylene ethyl acrylate)	$-[CH_2 - CH_2 - CH - CH_2 - CH_2]_n^-$ with branch $\overset{C}{\underset{O\,OC_2H_5}{}}$	10-24%

In a process similar to making popcorn or expanded snacks, readily degradable starch is used as an alternative to a conventional polymer. Selection of the type of starch is critical to achieving properties similar to expanded polystyrene (EPS) (*10*) and a small amount of plasticizing agent is required. The resulting material is water dispersible and biodegradable but has not yet reached the very low bulk densities achieved by EPS. There is also some concern that these resulting "peanuts" could attract rodents. Using technology from the USDA in Peoria, Unistar have attempted to commercialize a starch graft copolymer.

When we turn from the natural biodegradable polymers to synthetic biodegradable polymers, we are faced with an enormous variety of potential materials but so far only few have been commercialized, although considerable development is being carried out on polyester systems. The two most common polyesters being commercialized are polylactic acid and polycaprolactone, the former by e.g. Dupont/ConAgra in a venture called Ecochem, and the latter, among others, by Union Carbide and Solvay (Table I). The structure of the polymers are shown in Table III.

The vinyl alcohols (including PVOH) can be biodegradable depending on the amount of substitution and the molecular weight. These materials are proposed for use in conjunction with the destructurized starch to give improved properties, over the starc-based polymer alone.

Table III. Synthetic Biodegradable Polymers - Commercialized

Polylactide: $HO_2C \left[CH - O \overset{\overset{\displaystyle O}{\|}}{-C} \right]_n CH - OH$ (with CH_3 substituents)

Polycaprolactone: $H - O \left[(CH_2)_5 - \overset{\overset{\displaystyle O}{\|}}{C} - O \right]_n R \left[O - \overset{\overset{\displaystyle O}{\|}}{C} (CH_2)_5 \right]_n O - H$

The particular quality of lactic acid for polylactide is produced by a conventional fermentation process and it is then polymerized. Work on this process was also carried out at the Battelle Institute in Cincinnati.

In the process pioneered by Zeneca (formerly ICI), Biopol, a polyhydroxybutyrate and polyhydroxyvalerate blend is microbiologically produced. The structures of the two polymers are shown in Table IV. While the product has found commercial applications in cosmetic bottles, its extensive use is restricted by its processibility and price.

Table IV. Structure of Polyhydroxybutyrate and Valerate.

Polyhydroxybutyrate:

$$\left[\begin{array}{c} O \quad\quad CH_3 \\ \| \quad\quad\ | \\ C\text{-}CH_2\ CH\text{-}O \end{array} \right]_n$$

Polyhydroxyvalerate:

$$\left[\begin{array}{c} \quad\quad\quad CH_3 \\ \quad\quad\quad CH_2 \\ O \quad\quad | \\ " \quad\quad\ | \\ C\text{-}CH_2\ CH\text{-}O \end{array} \right]_n$$

Particulate Starch Systems

The other approach to producing a functional degradable plastic is to take a conventional thermoplastic polymer such as polyethylene and add a biodegradable substance (either natural or synthetic), and other additives to assist in the breakdown of the polymer. This approach was pioneered by Griffin in the 1970's (*11*). The first difficulty was the problem of the interaction between the hydrophilic starch granules and the hydrophobic polymer. Griffin proposed a simple, cost effective solution of coating the starch granules with a surface-active polymer, such as a silane. Other proposals suggest the use of a polar polymer as outlined in Table I, which can be effective in this technology as well as those utilizing gelatinized or partially gelatinized starch. Recent work has also indicated that other polymers such as maleic-anhydride modified polypropylene (*12*) or polyacrylic acid-modified polyethylene (*13*) may be effective at retrieving some of the properties lost by incorporating an inert hydrophilic filler. The silane treatment is used by Ecostar International and the reduction in properties is shown in Table V.

The original Griffin system relied upon the incorporation of an unsaturated acid or ester to help degrade the thermoplastic polymer. Chance encounters with transition metals in soil or compost could accelerate the oxidative breakdown of the polymer supposedly via the formation of peroxides. Newer

Table V. Effect on Physical Properties of LLDPE Film of Various Additions of
Surface-Modified Granular Starch Expressed as % of Value without Starch.

Starch %		0.0	3.0	6.0	9.0
Gauge (microns)		56	56	53	55
Elmendorf Tear	MD	100	109	110	138
	TD	100	94	95	95
Tensile at Yield	MD	100	103	102	103
	TD	100	98	97	102
Tensile at Break	MD	100	96	97	75
	TD	100	80	78	69
Elongation	MD	100	89	91	83
	TD	100	87	86	79
Dart Impact		100	83	91	61

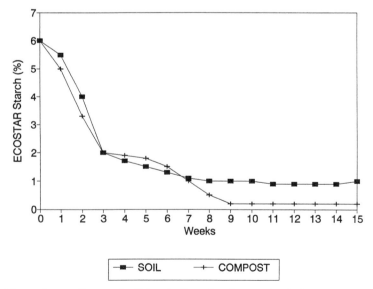

Figure 1.   Removal of Starch Granules from Polyethylene Film in
Various Environments

technology incorporates prodegradants with the other ingredients in a masterbatch and this is the basis of the ECOSTARplus® system.

ECOSTARplus® System

The particular features that the ECOSTARplus® system is designed to achieve are:

- processibility on conventional plastics machinery
- functionality similar to that of the conventional plastic article
- adequate stable shelf-life
- disintegration within a prescribed time frame
- ultimate mineralization
- no toxic effects
- small additional cost compared with the conventional plastic

Degradation of Plastics Containing ECOSTARplus®

There has been much debate about the mechanism of degradation and its measurement. However, over recent years research has provided a much better understanding of the process.

There are three distinct but related mechanisms that take place during environmental degradation of plastic articles containing ECOSTARplus®.

Table V. Mechanism of Degradation of Plastic Articles Containing ECOSTARplus®.

1. Digestion of the starch out of the plastic article.

2. Thermal oxidative breakdown or photodegradation of the polymer.

3. Digestion of the polymer fragments.

In the laboratory these mechanisms can be studied independently, but in natural environments they will take place contemporarily and there is synergy between all three.

Digestion of Starch from Plastic Articles

The ability of microorganisms to digest particulate starch from plastic articles has been shown by numerous investigators and it was examined quantitatively using FTIR by Ianotti and coworkers *(14)*. Their results are illustrated in Figure 1.

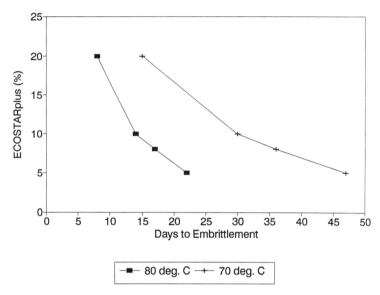

Figure 2.   Days to Embrittlement (Defined as Less than 5% of the
               Original Elongation) of LDPE Films

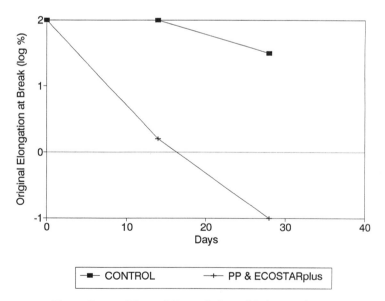

Figure 3.   Thermal Degradation of Polypropylene

The ability of extracellular enzymes to digest the starch entrapped in the polyethylene matrix was at first surprising, but Griffin showed that enzymes could penetrate thin membranes of polyethylene (2). The effect of this starch digestion is to weaken the continuous thermoplastic polymer matrix and to provide more surface area for the second mechanism.

Thermal Oxidative Breakdown and Photodegradation of the Polymer

ECOSTARplus® contains a sophisticated additive package to accelerate the oxidative and/or photodegradative breakdown of the polymer. This can be followed by measuring one or more of several parameters.

Table VI. Parameters used to measure thermal oxidation breakdown or photodegradation.

- Loss of elongation.
- Carbonyl group formation.
- Reduction of molecular weight.

This breakdown is well known in the plastic industry, but is accelerated with the ECOSTARplus® system. Graphs showing loss of elongation on thermal treatment of three different polymers are shown in Figure 2, 3 and 4. A temperature of 80°C is generally chosen to achieve faster comparative results.

Use of different polymers can have a dramatic effect on the rate of this degradation as is shown in Figure 5. It is not just the polymer that affects this rate but also the additives, particularly the antioxidants in the polymers that can modify the rate of breakdown as shown in Figure 6.

The degradation of the polymer can be measured by following the formation of carbonyl groups, which is related to the loss of elongation as shown in Figure 7. It is important to note the continued formation of carbonyl groups after the film is embrittled.

The reduction in molecular weight for LDPE is shown in Figure 8 and for HDPE in Figure 9. The delayed onset of the breakdown of the HDPE film is due to the presence of the necessary antioxidants.

Since this thermo-oxidative breakdown is a chemical process the question may be asked whether the starch is necessary! It has been shown that the presence of starch can affect the thermo-oxidative breakdown in an abiotic environment as is illustrated in Figure 10. The reason for this is not fully understood; it may be related to the chemical effect of the starch hydroxyl groups, but more likely is due to greater oxygen permeability of films containing starch.

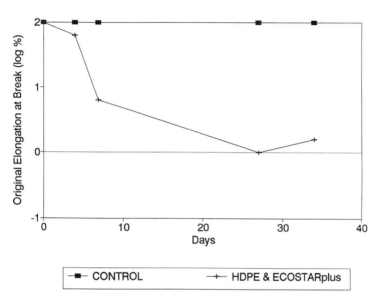

Figure 4.   Thermal Degradation of High Density Polyethylene

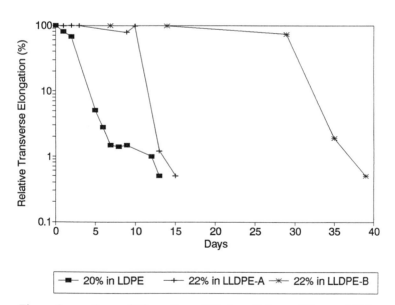

Figure 5.   Loss of Elongation of Various Polymer Films at 80°C

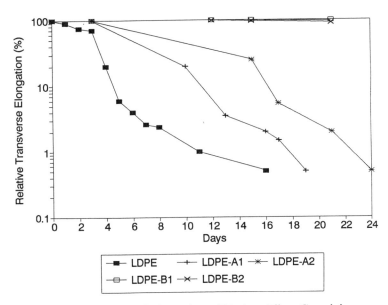

Figure 6.   Loss of Elongation of Various Films Containing Ecostarplus with Different Antioxidants.

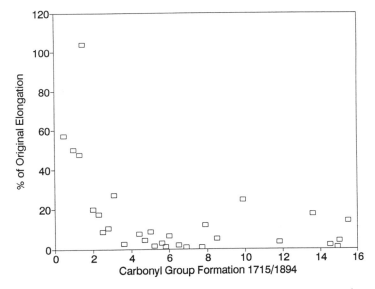

Figure 7.   Comparison of Carbonyl Group Formation (Measured as Area Under 1715 cm^{-1} Peak) with Loss of Elongation of Films Containing Ecostarplus.

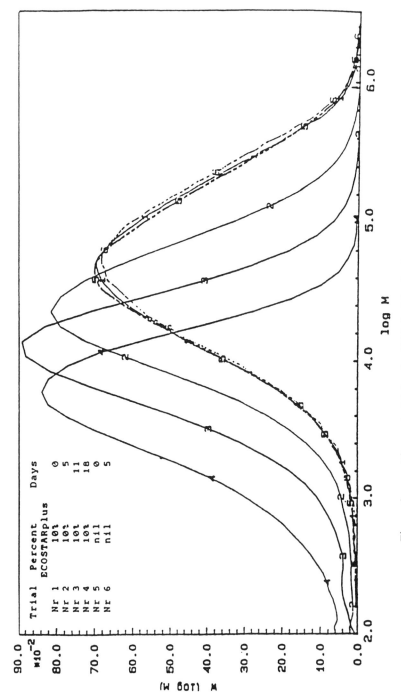

Figure 8.   Change in Molecular Weight Distribution of LDPE Films
            Containing Ecostarplus

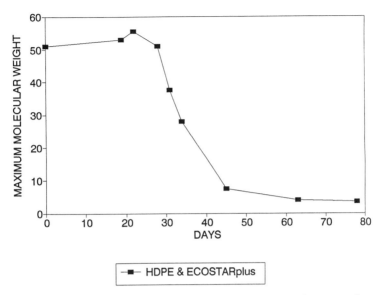

Figure 9.   Change in Maximum Molecular Weight of HDPE Films
             Containing 10% Ecostarplus

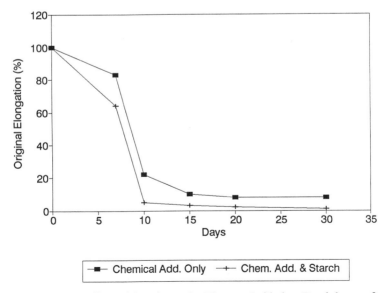

Figure 10.   Effect of Starch on the Thermo-Oxidative Breakdown of
             LDPE Film Containing Degradant Chemicals

More important, however, is the effect on the thermo-oxidative mechanism of digestion out the film of the starch. This has been investigated by using microtransmittance FTIR, and looking at carbonyl absorption in the region of a void formed by biological removal of a starch granule (Figure 11).

Digestion of the Polymer Fragments

Professor Albertsson at the Royal Institute of Technology in Stockholm, Sweden, has shown the very slow evolution of carbon dioxide from LDPE during soil burial *(15)*. The mechanism was assumed to involve first a chemical breakdown followed by microbiological attack on the polymer fragments.

Potts showed that low molecular weight fragments from polyethylene could support fungal growth *(16)*. While it is an oversimplification to state a simple molecular weight as the upper limit on microbiological digestion since it will depend on:

- chain branching
- presence of polar groups
- hydrophobicity of the surface.

However, it is generally accepted that low molecular weight hydrocarbon fragments can be digested by microorganisms.

A modified Stürm test was carried out using thermally oxidized high density polyethylene containing ECOSTARplus®. By measuring the carbon dioxide evolved it was proven that, in the relatively short time of the test, that some of the polyethylene was metabolized (Figure 12).

Preliminary results from the State University of New York at Buffalo have confirmed this biodegradation using ^{14}C-labelled polyethylene.

Conclusions from Degradation Tests

The results from the various degradation tests confirm that in the appropriate environments these products will degrade and sufficient is known about the mechanism and factors influencing the rate to make some predictions about breakdown performance.

Summary

The field of degradable plastics is changing rapidly both because of commercial factors and developments in the technology. One particular technology has been highlighted in this paper, but the different requirements for particular applications will provide opportunities for several technical solutions.

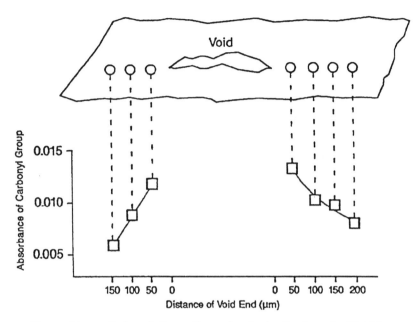

Figure 11.   Carbonyl Group Formation in the Region of a Void in a
Plastic Film Buried in Soil

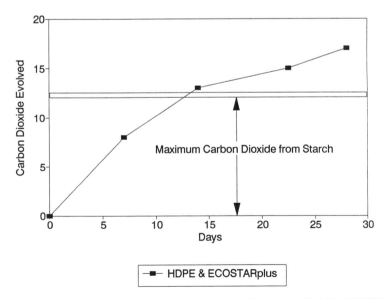

Figure 12.   Carbon Dioxide Evolution from Thermally Oxidized HDPE
Film Containing Ecostarplus

Literature Cited.

1.   Milstein, O.; Gersonde, R.; Hütterman, A.; Fründ, R.; Lüdermann, H.D.;
      Chen, M.J. and Meister, J.J., Bio/Environmentally Degradable Polymer
      Society 2nd Annual Meeting, Chicago, August, **1993**.

2.   Suzuki, T.; Appl, J.; Polymer Science: Applied Polymer Symposium,
      1979, 35, 431-437.

3.   Lindley, N.D.; Pedley, J.F.; Kay, S.P. and Heydeman, M.T.; International
      Biodeterioration, **1986**, Vol. 22 No. 4, 281.

4.   Buchanan, C.M.; Gardner, R.M.; Komarek, R.J.; Gedon, S.C. and White,
      A.W.; In Biodegradation of Cellulose Acetates in Biodegradable
      Polymers and Packaging; Ching, C.; Kaplan, D. and Thomas, E.,
      Lancaster, **1993**.

5.   US Patent 3,949,145, April 6, **1976**
      US Patent 4,133,784, January 9, **1979**.

6.   PCT Patent WO 91-02025, February 21, **1991**.

7.   U.S. Patent 4,673,438, June 16, **1987**.
      CHIMIA 41, 1987, Nr. 3 (März), p 76-81.

8.   US Patent 5,095,054, March 10, **1992**.

9.   Tomka, I.; Biologisch abbaubare Kunststoffe aus nachwachsenden
      Rohstoffen, Technische Akademie Esslingen Symposium Nr.
      14506/56.237, Novmeber **1991**.

10.   Altieri, P.A. and Lacourse, N.L.; Starch-Based Protective Loose Fill
      Material, Corn Utilization Conference IV, St. Louis, June **1992**.

11.   Griffin, G.J.L.; Biodegradable Fillers in Thermoplastics, ACS Advances
      in Chemistry, Series No. 134, **1975**.

12.   US Patent 5,026,745, June 25, **1991**.

13.   Adur, A.M.; and Constable, R.C.; Higher Performance
      Polypropylene/Starch Biodegradable Alloys, 4th International Congress
      on Compatibilizers and Reactive Polymer Alloying, New Orleans,
      February, **1991**.

14.   Ianotti, G.; Fair, N.; Tempesta, M.; Neibling, H.; Hsich, F.H.; Mueller, R.; In Degradable Materials Perspectives, Issues and Opportunities; Barenberg, S.A.; Brash, J.L.; Narayan, R.; and Redpath, A.E.; eds; CRC Press, Boca Raton, FL, **1990**, pp 425-439.

15.   Albertson, A.C.; J. Applied Polymer Science, **1988**, 35, 1289.

16.   Potts, J.; et al, Polymer Preprints, **1972**, 12(2), 629.

RECEIVED July 7, 1994

STARCH, STARCH BLENDS,
AND COMPOSITES

Chapter 3

Melt Rheology of Thermoplastic Starch

J. L. Willett, B. K. Jasberg, and C. L. Swanson

Plant Polymer Research Unit, National Center for Agricultural Utilization Research, Agricultural Research Service, U.S. Department of Agriculture, Peoria, IL 61604

The use of starch as a thermoplastic material is a recent development. An understanding of the rheology of thermoplastic starch melts is needed in order to understand the effects of processing on structure/property relationships. This article discusses the effects of temperature, moisture content, molecular weight reduction (hydrolysis), and low molecular weight additives on the behavior of thermoplastic starch melts. Thermoplastic starch melts exhibit power law behavior. The melt viscosity decreases with increasing temperature, moisture content, and decreasing molecular weight. Low molecular weight additives also reduce the viscosity. Glycerol monostearate slightly increases the melt viscosity. This effect is attributed to the formation of helical inclusion complexes which are stable at the extrusion temperatures. The power law index and the rate of change in viscosity with temperature of thermoplastic starch melts are similar to those of synthetic polymers.

Concern over the magnitude of discarded waste in the U. S. has in recent years stimulated a great deal of research in the area of degradable plastics (1,2). Several approaches have been developed to either render conventional plastics degradable after disposal, or to produce inherently degradable materials. In particular, starch, because of its abundance and low cost, has received much attention as a raw material in both of these approaches to degradability. Other advantages to the use of starch as a raw material are the fact that it is a renewable resource, its potential for replacing nonrenewable petroleum or natural gas feedstocks, and potential economic benefits to the agricultural industry by providing new markets for agricultural products (3).

The use of starch in plastics has progressed through several stages. Early attempts include efforts to polymerize allyl starch (4), the use of amylose to produce films (5-7), and studies of amylose acetate and starch acetate as competitors to cellulose acetate (8). Efforts in the early to mid 1970s focused on using granular starch as a filler in rubbers, PVC, and polyolefins (9-15). Granular starch formulations were generally limited to starch contents of approximately 10% by weight or less, due to the general deterioration of mechanical properties seen in filled materials (16). Blends of gelatinized starch with water soluble or water dispersible polymers were developed in the late 1970s and early 1980s (17-19). These materials had the advantage of starch contents up to 50-60% with good properties, but used relatively high cost raw materials. Another avenue of starch incorporation in plastics is via graft polymerization (20-22). Grafting of unsaturated monomers (e.g. methyl acrylate) onto the starch molecule backbone yields materials with starch contents of up to 60%. These graft copolymers can be injection molded or extruded into films with properties similar to low density polyethylene (23).

The extrusion of starch offers a route to materials with high starch content, relatively low raw materials cost, and inherent degradability. Although the food industry has been extruding starch for many years (24), only recently has extruded starch been considered as a thermoplastic material (25-27). The rheology of starch melts has therefore received little consideration from the polymer processing point of view.

Starch is a polysaccharide of repeating glucose units. It is a mixture of two polymers, amylose and amylopectin (7,28). Amylose is a predominantly linear, lightly branched polymer, comprised of (1->4) α-D linkages with number average molecular weights in the range of several hundred thousand. Amylopectin is highly branched, with intermittent (1->6) links; its molecular weight is typically several million, and can be as great as 50 million. Corn starch typically contains amylopectin and amylose at a ratio of approximately 3:1; genotypes with higher amylose contents (up to 70%) or virtually no amylose (waxy maize starch) are available commercially. Native corn starch granules have an average diameter of approximately 10 microns.

The structural differences in the two forms of starch have a considerable effect on the properties of starch materials. Because of its branched structure, amylopectin generally has inferior mechanical properties relative to amylose. Wolff et al (6) showed that increasing the amylopectin content in amylose/amylopectin films decreased the tensile strength and the elongation. Similar behavior has been observed with synthetic polymers such as polystyrene (29,30). Data of Lai and Kokini (31) show that at constant temperature, moisture content, and shear rate, high amylose starch melts have a higher viscosity than waxy maize starch melts. Similar branching effects have been observed for polyethylene (32) and polystyrene (29) melts. The shear strength of expanded starch products decreases with increasing amylopectin content (33).

Extrusion of Starch

The three hydroxyl groups on each monomer of the starch molecule lead to significant hydrogen bonding between adjacent molecules. Penetrant molecules which can disrupt this hydrogen bonding are able to disperse starch into solution; this process is generally referred to as gelatinization. The melting point of starch at atmospheric pressure and low moisture content, like that of cellulose (a (1->4) β-D glucose polymer), exceeds its decomposition temperature. In closed systems in the presence of water, the melting point of starch decreases with increasing moisture content in a manner described by the Flory model of diluent effects (34).

Starch can be gelatinized in the high temperature, high shear, and high pressure conditions found in extruders. The result is a melt of interspersed amylopectin and amylose molecules and destruction of the granule structure. The effects of extrusion conditions and various additives on starch as a food product have received considerable attention (35-47).

The melt viscosity of starch can be described by equations of the form

$$\eta = K\dot\gamma^{m-1}\exp(\frac{E_a}{RT})\exp(-\alpha MC) \tag{1}$$

where η is the viscosity, K is the consistency, T is the absolute temperature, MC is the moisture content, $\dot\gamma$ is the shear rate, m is the power law index, E_a is an activation energy, R is the gas constant, and α is the moisture content coefficient. This relationship was first suggested by Harper (48,49), and has been verified by several groups. Various modifications to this equation have been added to account for the effects of screw speed in twin screw extrusion (50), specific mechanical energy input (51), residence time on degradation kinetics (31), and the degree of cooking or starch conversion (31).

The effects of additives on starch melt viscosity have also been examined. Amylose in solution forms helical inclusion complexes with linear fatty acids and esters, as well as monoglycerides and some diglycerides. These materials form complexes with starch during extrusion (37,39,42), but little is known of the stability of the complexes during extrusion. Triglycerides and fats which do not complex with starch influence the rheology of molten starch by reducing the degree of molecular weight degradation (37,39). Most of the work in this area has focused on the influence of additives on extrudate properties such as water solubility and water absorption rather than rheological effects.

Various workers have shown that considerable molecular weight degradation occurs during the extrusion of starch (36,39,52-54). Size exclusion chromatography shows that the amylopectin undergoes more chain scission than amylose. Little if any low molecular weight oligosaccharides are formed during extrusion, suggesting that shear forces encountered during extrusion are more important than thermally controlled bond scission. Theories of mechanical degradation of polymers (55) predict that in the presence of shear, the largest molecules are the most likely to degrade, while few of the lower molecular weight chains will do so. The degradation of amylopectin observed during extrusion is consistent with this prediction.

The purpose of this paper is to describe the effects of temperature, moisture content, additives, and molecular weight on the melt viscosity of thermoplastic starch. The starch was pelletized prior to the viscosity measurements. This technique allowed the moisture content to be changed before the measurement step to elucidate the MC effects during pelletizing on viscosity. Formation and thermal stability of starch/additive complexes were investigated using X-ray diffraction and differential scanning calorimetry (DSC).

Experimental Details

Buffalo 3401 unmodified cornstarch from CPC International was used. Urea (EM Science), lecithin (Fisher Scientific), triethylene glycol (TEG, Fisher Scientific), glycerol monostearate (GMS, Kemester 5500, Witco), and polyoxyethylene stearate (POES, ICI Americas) were used as received. Stardri 1, a waxy maize derived dextrin (A. E. Staley), was used as a low molecular weight starch analog. Distilled water was used for all formulations.

Pelletizing and rheological measurements were performed using a Brabender PL 2000 torque rheometer. A 19 mm diameter, 40/1 L/D double mixing zone screw was used for pelletizing at a screw speed of 60 rpm. The starch melt was extruded through a 17-hole strand die (1.6 mm diameter holes) and air cooled. The temperature profile during pelletizing was 140/160/160/130°C (feed zone to die). The die zone temperature was reduced to 100°C when pelletizing starch at 30% MC to prevent foaming by steam expansion.

Viscosity measurements were performed with a 25/1 L/D, 19 mm diameter screw with a 3/1 compression ratio. Capillary dies were 2 mm in diameter, with length/radius (L/R) ratios 20/1, 30/1, and 40/1. Die temperatures for the starch/water compounds ranged from 110°C to 180°C. All additive formulations were pelletized with 20% MC and equilibrated to 15% MC for viscosity measurements. Additive/starch viscosities were measured with a temperature profile of 150/155/160/160°C. Melt temperature and pressure were measured with a combination thermocouple/transducer unit (Dynisco) in contact with the melt just prior to the die entrance.

Shear rates were determined using the Rabinowitch correction (56,57):

$$\dot{\gamma}_c = \dot{\gamma}_a \frac{(3b+1)}{4b} \tag{2}$$

$$\dot{\gamma}_a = \frac{4Q}{\pi R^3} \tag{3}$$

where $\dot{\gamma}_c$ is the corrected shear rate, $\dot{\gamma}_a$ is the apparent shear rate, and b is the slope of the plot of log(shear stress) versus log($\dot{\gamma}_a$). Q is the volumetric output, and R is the capillary radius. Values for the power law index m and consistency K were determined by linear regression of log(η) versus log($\dot{\gamma}_c$) plots.

X-ray diffraction patterns were obtained using a Phillips PW 1820 diffractometer with a slit opening of 0.20 mm. Readings were taken at intervals of 0.05 2θ units at a rate of 4 scans per second.

Melting curves of starch/additive extrudates were obtained on a Perkin-Elmer Series 7 Thermal Analysis System. The scan rate was 10°C/minute over the range 5°C to 220°C. Sample weight was approximately 30 mg, and the moisture content was 15%.

Results

The effect of temperature on melt viscosity of starch with 30% MC (pelletized with 20% MC) is shown in Figure 1. As with other thermoplastic materials, the melt viscosity decreases with increasing temperature. This trend was observed regardless of the MC during either the pelletizing step or the viscosity measurement, although the effect was more pronounced with starch pelletized at 20% and 30% MC.

The power law index m increased with melt temperature at constant MC. For the data in Figure 1, m values were 0.31 ($\pm$ 0.04) at 110°C, 0.34 ($\pm$ 0.03) at 120°C, and 0.40 ($\pm$ 0.03) at 130°C. m values for starch pelletized with 30% MC and tested at 30% MC increased from 0.31 ($\pm$ 0.03) at 110°C to 0.37 ($\pm$ 0.03) at 130°C. These results are consistent with those of Lai and Kokini (31), who studied the melt viscosity of waxy maize starch and high amylose starch. This increase in m indicates that the melt is more Newtonian at higher temperatures.

Figure 2 shows the effect of moisture content on starch melt viscosity at 130°C for starch pelletized with 15% MC. As predicted by Equation 1, the viscosity decreases as the MC increases. Similar behavior is observed for starch pelletized and tested at different moisture contents and temperatures.

Increasing the moisture content decreases the power law index. For starch pelletized with 15% MC, m decreased from 0.49 to 0.32 at 160°C when the MC was increased from 15% to 30%. This result is consistent with the behavior of waxy maize and high amylose starches (31), and appears to be a general feature of starch melts. A larger decrease in m is seen for waxy maize starch than for high amylose starch over the same range of MC (31).

Increased moisture content reduces the melt viscosity and hence the shear stresses acting on the starch molecules during extrusion. Since shear forces dominate the chain scission process rather than temperature effects during extrusion, a reduction in shear stress due to increased MC should lead to reduced molecular degradation. It is therefore reasonable to expect that increased moisture content during extrusion lowers the degree of chain scission. Since the higher molecular weight starch is less Newtonian (see dextrin results below), its power law index is lower.

The moisture content during the pelletizing step also affects the viscosity. This effect is seen when starches pelletized at different moisture contents are equilibrated to equal moisture contents before the viscosity measurement. Figure 3 illustrates this effect when the pelletized starch is equilibrated to 30% MC after pelletizing.

The separation of the three viscosity curves shows the effect of moisture content during pelletizing on subsequent viscosity. However, the increase in viscosity with MC is not monotonic, but exhibits a maximum at a pelletizing moisture content of 20%. One possible explanation for this result is the interaction between higher shear stresses at lower MC and increased hydrolytic chain scission due to the greater residence time at higher MC (53).

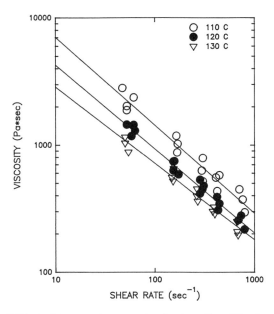

Figure 1. Effect of temperature on starch viscosity. Starch pelletized at 20% MC; viscosity measured at 30% MC.

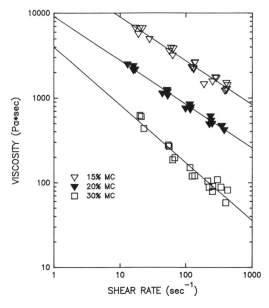

Figure 2. Effect of moisture content on starch at 160°C. Starch pelletized at 15% MC.

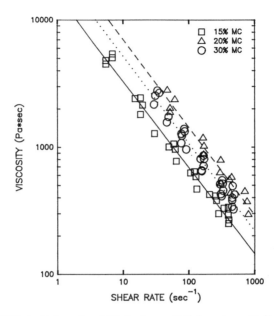

Figure 3. Effect of changing MC during pelletizing step. Pelletized starch equilibrated to 30% MC before viscosity measurement.

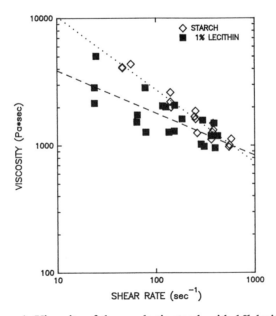

Figure 4. Viscosity of thermoplastic starch with 1% lecithin.

Low molecular weight additives also affect the viscosity of thermoplastic starch melts. Figures 4-7 show the effects of various additives.

Lecithin and POES significantly reduce the viscosity of starch at 15% MC when added at a 1% by weight level, as shown in Figures 4 and 5. Both additives lower the power law index relative to starch/water melts. Therefore, the additive effects are more pronounced at lower shear rates than at higher ones. It has been shown that adding lecithin during twin screw extrusion of starch yields extrudates with higher intrinsic viscosity, suggesting a reduction in chain scission (39). It therefore seems that lecithin and POES reduce m by retaining a higher molecular weight of the starch during extrusion. At levels greater than 1%, both lecithin and POES overlubricated the melts to the extent that continuous extrusion was not possible due to excessive slip. The amount of scatter in Figure 4 suggests that even 1% lecithin is approaching the limit of overlubrication.

The effects of urea and TEG on starch melt viscosity are shown in Figures 6 and 7. These additives also reduce the melt viscosity, but not as effectively as lecithin and POES. A comparison of Figures 5 and 7 indicates that 5% TEG is needed to reduce the viscosity as effectively as 1% POES. The scatter in the 5% TEG data suggest this formulation is approaching overlubrication; formulations with more than 5% TEG could not be continuously extruded.

Urea is similar to TEG in ability to reduce starch viscosity at a 2% level. When the urea is increased to 5%, a reduction in viscosity is seen, but it is not as great as might be expected in comparison to TEG. The effectiveness of urea as a plasticizer may be diminished if it is partially hydrolyzed to CO_2 and NH_3 during extrusion. Samples of starch/urea extrudates stored in sealed bags had an ammonia odor when opened, suggesting hydrolysis did occur during extrusion.

GMS has an unusual effect on starch melt viscosity, as shown in Figure 8. The addition of 2% GMS reduces the consistency by less than 10%, a lesser effect than the other additives. The consistency of 5% GMS is about 30% greater than the starch/water control. GMS also reduces the power law index slightly. Compared to the behavior of the other additives in this study, GMS clearly has a different type of effect on starch melt viscosity.

The values of m and K for the additive/starch systems are summarized in Table 1.

Molecular weight also influences melt viscosity. A dextrin was used as a low molecular weight analog of starch to examine the effect of molecular weight. Dextrins are the products of hydrolytic degradation of starch. Therefore, the use of dextrin yields qualitative information on the influence of molecular weight, even though the molecular weights of the materials used in this study were not directly measured. Dextrin was extruded with 15% MC at 110°C, 120°C, and 130°C, and with 20% MC at 100°C, 110°C, and 120°C. The effect of temperature on dextrin viscosity with 15% MC is shown in Figure 9a. As with starch, increasing the melt temperature lowers the melt viscosity and increases the power law index. The power law index change is much more pronounced than with starch. At 130°C, the power law index is 0.94, indicating a highly Newtonian fluid.

Similar behavior is observed when the moisture content is increased to 20% at extrusion temperatures of 100°C, 110°C, and 120°C, as shown in figure 9b.

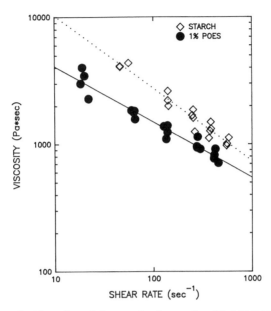

Figure 5. Viscosity of thermoplastic starch with 1% POES.

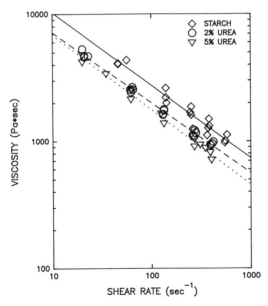

Figure 6. Viscosity of thermoplastic starch with 2% and 5% urea.

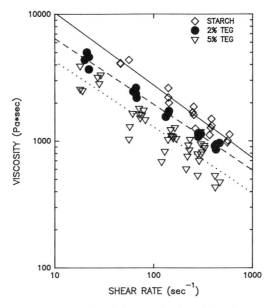

Figure 7. Viscosity of thermoplastic with 2% TEG.

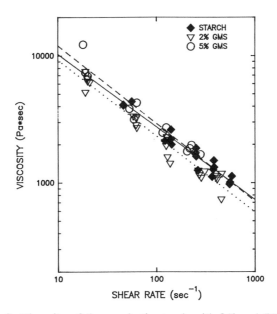

Figure 8. Viscosity of thermoplastic starch with 2% and 5% GMS.

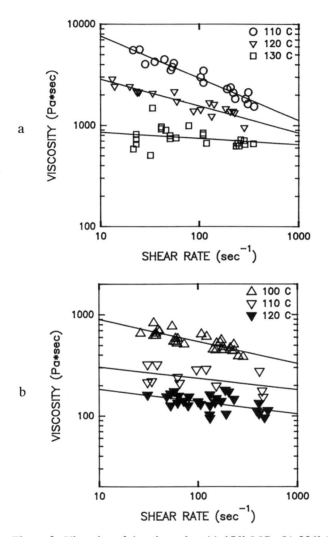

Figure 9. Viscosity of dextrin melts: (a) 15% MC; (b) 20% MC.

The power law index increases from 0.78 at 100°C to 0.86 at 110°C. Increasing the temperature to 120°C increases m to 0.88.

Higher moisture contents reduce the viscosity of the dextrin, as shown in Figures 10a and 10b. The melts are also more Newtonian, as evidenced by an increase in m with moisture content at both temperatures.

Discussion

The starch melts in this study exhibited power law behavior. In a power law fluid, the shear stress τ is dependent on the shear rate in the following manner (56,57):

$$\tau = K\dot{\gamma}^m \tag{4}$$

where m and K have the same definitions as in Equation 1. Substituting the above relationship into the Newtonian definition of the viscosity $\eta = \tau/\dot{\gamma}$, one has for a power law fluid

$$\eta = K\dot{\gamma}^{m-1} \tag{5}$$

For shear thinning melts such as thermoplastics, $0 < m < 1$. The smaller the value of m, the more shear thinning the melt. Conversely, as m approaches 1, the melt becomes more Newtonian and exhibits less shear thinning. Values of the power law index for thermoplastic starch melts in this study ranged from 0.31 to 0.50, depending on the moisture content and temperature. When the molecular weight is significantly reduced, the melts exhibit an increasing degree of Newtonian behavior, indicated by m values greater than 0.80 for the dextrin.

The m values of thermoplastic starch are similar to those of polyethylene, polystyrene, and poly(methyl methacrylate) (56,57). Thermoplastic starch melts are therefore similar in character to synthetic polymer melts with regards to shear thinning. Changes in the power law index of thermoplastic starch with temperature are similar to those observed for synthetic thermoplastics (57).

Thermoplastic starch melts are similar to melts of synthetic polymers with low glass transition temperatures. A measure of the temperature sensitivity of a melt is given by the relative change in viscosity with temperature (56):

$$\frac{-1}{\eta(T_r)}\frac{\partial\eta}{\partial T} \approx \frac{-1}{\eta(T_r)}\frac{(\eta-\eta(T_r))}{(T-T_r)} \tag{6}$$

where T_r is the reference temperature. Typical values of this parameter for the starch melts in this study range from 0.026 to 0.033. By comparison, values for polyethylene, polyamide-6, polyamide-6,6, polypropylene, and poly(ethylene terephthalate) are in the range of 0.02 to 0.03. Polymers with higher T_gs such as poly(styrene), poly(methyl methacrylate), and styrene-acrylonitrile copolymers have values of 0.10 to 0.20 (56).

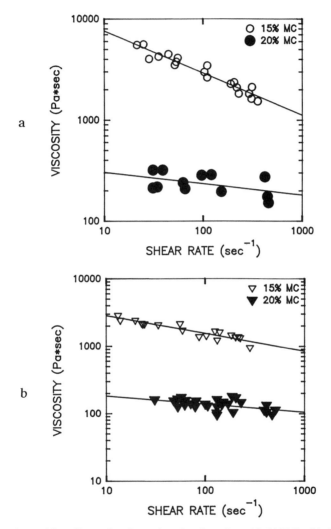

Figure 10. Effect of MC on dextrin viscosity: (a) 110°C; (b) 120°C.

Reduced activation energies (E_a/R) can be determined at constant moisture content from the slope of ln(K) versus 1/T plots, where T is in Kelvins. Values for the thermoplastic starch melts in this study are 9950, 8500, and 8500 Kelvins at moisture contents of 15%, 20%, and 30% respectively. These value are higher than those reported by others (51), and lower than the value reported for potato starch (25).

Activation energies for the dextrins are much higher than those of thermoplastic starch. At 15% MC, E_a/R is 23,300 Kelvins, decreasing to 13,300 at 20% MC. The activation energy for dextrin is not simply a measure of flow activation, but also reflects the changes in power law index with temperature. This dependence of m on temperature increases E_a/R relative to systems where m is less sensitive to temperature.

The moisture content coefficient α is evaluated from the slope of ln (K) versus MC plots at constant temperature. At 160 °C, a value of 12.6 is calculated, which is in agreement with those of other workers under similar conditions (31,51). Values of α for dextrin are considerably higher: 78.4 at 120°C and 61.6 at 110°C. The decrease in α with increasing temperature is consistent with the reported behavior of high amylose and waxy maize starches (31).

All of the additives in this study reduced the melt viscosity, except GMS. X-ray diffraction patterns for starch extruded with 2% GMS, 1% POES, and 1% lecithin are shown in Figure 11 with patterns from extruded starch and native starch. It is clear that the crystalline structure of the native starch is disrupted by the initial extrusion. The diffraction peaks at 2θ values of 13.1 and 20.1 indicate that these three additives form helical inclusion complexes with amylose. The small peaks in the extruded starch sample at these 2θ angles are consistent with complex formation due to the small amount of fatty acid in the native starch. POES appears to be better at complex formation than lecithin, which is consistent with the fact that POES is linear and lecithin is a diglyceride. GMS and POES are approximately equal in ability to complex, since the peak heights are approximately 2:1 (2% GMS versus 1% POES). The diffraction patterns were measured on extrudates at room temperature, and therefore contain no information regarding the relative thermal stabilities of the various complexes.

The thermal stability of the complexes was investigated using DSC. Samples with 15% MC were heated at a rate of 10°C/minute up to 220°C. All three samples exhibited broad exotherms above 125°C. Exotherm peak temperatures were in the order lecithin < POES < GMS, with the GMS peak maximum at approximately 210°C. Since the melting point of amylose at 15% MC is approximately 250°C (58), these exotherms may be due to recrystallization of amylose released by the melting of the helical inclusion complexes. The DSC results suggest that the GMS/amylose complexes have greater thermal stability, as measured by exotherm peak temperature, than POES or lecithin complexes, and may be stable under the extrusion conditions during viscosity measurement.

Using the results of Karkalas and Raphaelides (59), one can estimate that 2.3% GMS by weight is needed to complex all of the available amylose in corn starch, assuming complete complex formation. Therefore, at GMS levels below 2.3%, there is no excess GMS available to lubricate the melt. One would expect the helical complexes to be relatively rigid in the melt and hence increase the

TABLE 1.   EFFECTS OF ADDITIVES ON THERMOPLASTIC STARCH
MELT VISCOSITY[a]

ADDITIVE	WT. PER CENT	CONSISTENCY (Pa*sec^m)	m
Control	---	38,900 (3,800)	0.43 (0.03)
TEG	2	21,400 (1,500)	0.48 (0.02)
TEG	5	14,100 (3,600)	0.48 (0.04)
Urea	2	25,700 (1,200)	0.45 (0.01)
Urea	5	26,900 (1,900)	0.41 (0.02)
GMS	2	36,300 (5,400)	0.40 (0.03)
GMS	5	50,100 (8,800)	0.39 (0.05)
Lecithin	1	8,300 (2,600)	0.67 (0.07)
POES	1	11,220 (1,400)	0.56 (0.02)

[a] Numbers in parentheses are standard deviations.

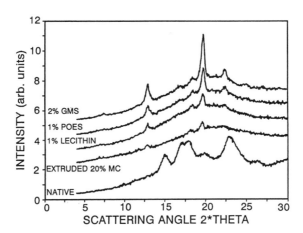

Figure 11. X-ray diffraction patterns of native starch and extruded starch
with and without additives: (1) native starch; (2) extruded with 20% MC;
(3) 1% lecithin; (4) 1% POES; (5) 2% GMS.

viscosity.   If the GMS/amylose complexes are stable under the experimental extrusion conditions (15% MC and 160°C), the 2% GMS melt should have a higher viscosity than the 5% GMS melt, since there would be excess GMS in the latter case.   Figure 8 shows the opposite to be the case.   Although it has been shown that amylopectin does not complex with emulsifiers such as GMS in solution (60), little is known about amylopectin complex formation in low moisture systems.   The low moisture conditions during extrusion may drive the GMS/amylopectin complex equilibrium to the right compared to solution behavior. Further investigations of these systems are in progress.

If POES or lecithin complexes were stable during extrusion, a viscosity increase with additive level would be expected, based on a similar calculation. The fact that levels of POES or lecithin above approximately 1% do not extrude well due to overlubrication suggests that complexes of these additives with amylose are not stable during extrusion.   Therefore, the viscosity and DSC data suggest that GMS/amylose complexes are stable under extrusion conditions of 15% MC and 160°C, and are responsible for the relatively high viscosities observed for these melts.   This effect could play an important role in the role of processing on structure/property relationships and morphology of extruded starch.

The melt viscosities of thermoplastic starch (15% and 20% MC) and a low density polyethylene (LDPE) at 160°C are shown in Figure 12.   The LDPE is Petrothene 3404B from Quantum Chemical; its melt index is 1.8 grams/10 minutes (at 190°C).   The viscosity of the LDPE falls between the two starch melt viscosities.   The LDPE is more shear thinning than the starch melts over the shear rate range studied.   It is clear from this figure that thermoplastic starches can be formulated to provide viscosities and shear thinning characteristics similar to those of commercially available commodity plastics in use today.

Conclusions

The effects of moisture content, temperature, molecular weight reduction (hydrolysis), and additives on melt viscosity of thermoplastic starch have been studied.   Thermoplastic starch exhibits power law behavior, with the power law index m being a function of temperature and moisture content.   m increased with increasing temperature; increasing the MC decreased m.   Dextrin had m values approaching 1, indicating more Newtonian behavior.   Starch pelletized with 20% MC had a higher viscosity than starch pelletized at either 15% MC or 30% MC, after equilibration to equal moisture contents.   Moisture content affected melt viscosity in an exponential fashion.   All the additives studied reduced the melt viscosity with the exception of GMS; lecithin and POES were the most effective. GMS at 2% loading had little effect on melt viscosity; 5% GMS slightly increased the viscosity.   This behavior may be due to unmelted GMS/amylose helical inclusion complexes which are stable at the extrusion conditions.   At 160°C, thermoplastic starch had a melt viscosity similar to an LDPE with a melt index of 1.8.

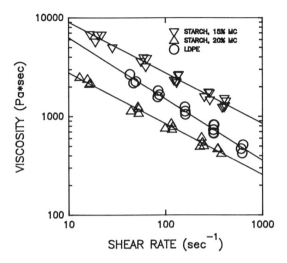

Figure 12. Comparison of viscosities of thermoplastic starch and LDPE melts at 160°C.

Acknowledgements

The authors gratefully acknowledge the efforts of R. P. Westhoff, R. Haig, and G. Grose in pelletizing and viscosity measurements, and the assistance of Dr. R. S. Shogren in the X-ray diffraction and DSC analysis.
The mention of firm names or trade names does not imply that they are endorsed or recommended by the U. S. Department of Agriculture over other firms or similar products not mentioned.

Literature Cited

1. *Proceedings of the Symposium on Degradable Plastics*; Society of the Plastics Industry: Washington, D.C., 1987.
2. *Degradable Materials: Perspectives, Issues and Opportunities*; Barenberg, S. A.; Brash, J. L.; Narayan, R.; Redpath, A. E., Eds.; CRC Press: Boca Raton, FL, 1990.
3. Doane, W.M.; Swanson, C.L.; Fanta, G.F. In *Emerging Technologies for Materials and Chemicals from Biomass*; Rowell, R. M.; Schultz, T. P.; Narayan, R., Eds.; American Chemical Society, 1992; 198-230.
4. Wilham, C. A.; McGuire, T. A.; Rudolphi, A. S.; Mehltretter, C. L. *J. Appl. Polym. Sci.* **1963**, *7*, 1403.
5. Lloyd, N. E.; Kirst, L. C. *Cereal Chemistry* **1963**, *40*, 154.
6. Wolff, I. A.; Davis, H. A.; Cluskey, J. E.; Gundrum, L. J.; Rist, C. E. *Ind. Eng. Chem. Res.* **1951**, *43(4)*, 915.
7. Young, A.H. Fractionation of Starch, in *Starch Chemistry and Technology*; Whistler, R.L.; BeMiller, J.N.; Paschall, E.F., Eds.; Academic Press, Orlando, FL, 1984; 249-284.

8. Rutenberg, M.W.; Solarek, D. Starch Derivatives: Production and Uses, in *Starch Chemistry and Technology*; Whistler, R.L.; BeMiller, J.N.; Paschall, E.F., Eds.; Academic Press, Orlando, FL, 1984; 312-388.
9. Katz, H. C.; Kwolek, W. F.; Buchanan, R. A.; Doane, W. M.; Russell, C. R. *Starch* **1974**, *26*, 201.
10. Katz, H. C.; Kwolek, W. F.; Buchanan, R. A.; Doane, W. M.; Russell, C. R. *Starch* **1976**, *28*, 211.
11. Buchanan, R. A.; Kwolek, W. F.; Katz, H. C.; Russell, C. R. *Starch* **1971**, *23*, 350.
12. Buchanan, R. A. *Starch* **1974**, *26*, 165.
13. G. J. L. Griffin, *U. S. Patent* **4,016,117**, (1977).
14. G. J. L. Griffin, *U. S. Patent* **4,021,388**, (1977).
15. G. J. L. Griffin, *U. S. Patent* **4,125,495**, (1978).
16. Nielsen, L.E. *Mechanical Properties of Polymers and Composites*; Marcel Dekker: New York, NY, 1974; 386-414.
17. Otey, F. H.; Mark, A. M.; Mehltretter, C. L.; Russell, C. R. *Ind. Eng. Chem. ,Prod. Res. Dev.* **1974**, *13*, 90.
18. Otey, F. H.; Westhoff, R. P.; Russell, C. R. *Ind. Eng. Chem. ,Prod. Res. Dev.* **1977**, *16*, 305.
19. Otey, F. H.; Westhoff, R. P.; Doane, W. M. *Ind. Eng. Chem. ,Prod. Res. Dev.* **1980**, *19*, 592.
20. Bagley, E. B.; Fanta, G. F.; Burr, R. C.; Doane, W. M.; Russell, C. R. *Polym. Eng. Sci.* **1977**, *17*, 311.
21. Bagley, E.B.; Fanta, G.F. Starch Graft Copolymers In *Encyclopedia of Polymer Science and Technology*, Mark, H.F.; Bikales, N.M., Eds.; John Wiley and Sons: New York, NY, 1977, 665-699.
22. Fanta, G. F.; Burr, R. C.; Doane, W. M.; Russell, C. R. *J. Appl. Polym. Sci.* **1977**, *21*, 425.
23. Dennenberg, R. J.; Bothast, R. J.; Abbott, T. P. *J. Appl. Polym. Sci.* **1978**, *22*, 459.
24. Mercier, C.; Linko, P.; Harper, J.M. *Extrusion Cooking*; American Association of Cereal Chemists: St. Paul, MN, 1989.
25. Tomka, I. Thermoplastic Starch In *Water Relationships in Food*; Levine, H.; Slade, L., Eds.; Plenum Press, New York, NY, 1991; 627-637.
26. Lay, G.; Rehm, J.; Stepto, R. F.; Thome, M.; Sachetto, J. P.; Lentz, D. J.; Silbiger, J. *U. S. Patent* **5,095,054**, (1992).
27. Wiedmann, W.; Strobel, E. *Starch* **1991**, *43*, 138.
28. Whistler, R.L.; Daniel, J.R. Molecular Structure of Starch. In *Starch Chemistry and Technology*, Whistler, R.L.; BeMiller, J.N.; Paschall, E.F., Eds.; Academic Press: Orlando, FL, 1984; 153-183.
29. Toggenburger, R.; Newman, S.; Trementozzi, Q. A. *J. Appl. Polym. Sci.* **1967**, *11*, 103.
30. Wyman, D. P.; Eylash, L. J.; Frazer, W. J. *J. Polym. Sci.* **1965**, *3*, 681.
31. Lai, L. S.; Kokini, J. L. *J. Rheo.* **1990**, *34*, 1245.
32. Busse, W. F.; Longworth, R. *J. Polym. Sci.* **1962**, *58*, 49.
33. Chinnaswamy, R.; Hanna, M. A. *Cereal Chemistry* **1988**, *65*, 138.
34. Donovan, J. W. *Biopolymers* **1979**, *18*, 263.

35. Anderson, R. A.; Conway, H. F.; Pfeifer, V. F.; Griffin, E. L. *Cereal Science Today* **1969**, *14*, 4.
36. Mercier, C.; Feillet, P. *Cereal Chemistry* **1975**, *52*, 283.
37. Mercier, C.; Charbonniere, R.; Grebaut, J.; Gueriviere, F. *Cereal Chemistry* **1980**, *57*, 4.
38. Paton, D.; Spratt, W. A. *Cereal Chemistry* **1981**, *58*, 216.
39. Colonna, P.; Mercier, C. *Carbohydrate Polymers* **1983**, *3*, 87. 40. Colonna, P.; Melcion, J. P.; Vergnes, B.; Mercier, C. *J. Cer. Sci.* **1983**, *1*, 115.
41. Gomez, M. H.; Aguilera, J. M. *Journal of Food Science* **1983**, *48*, 378.
42. Launay, B.; Lisch, J.M. Twin-screw Extrusion Cooking of Starches: Flow Behavior of Starch Pastes, Expansion and Mechanical Properties of Extrudates In: *Extrusion Cooking Technology*; Jowitt, R., Ed.; Elsevier Applied Science: London, 1983; 159-180.
43. Owusu-Ansah, J.; Voort, F. R.; Stanley, D. W. *Cereal Chemistry* **1983**, *60 (4)*, 319.
44. Colonna, P.; Doublier, J. L.; Melcion, J. P.; Monredon, F.; Mercier, C. *Cereal Chemistry* **1984**, *61 (6)*, 538.
45. Chinnaswamy, R.; Hanna, M. A. *Cereal Chemistry* **1990**, *67(5)*, 490.
46. Erdemir, M. M.; Edwards, R. H.; McCarthy, K. L. *Food Science and Technology* **1992**, *25*, 502.
47. Smith, A.C. Studies on the Physical Structure of Starch-Based Materials in the Extrusion Cooking Process In *Food Extrusion Science and Technology*; Kokini, J.L.; Ho, C.-T.; Karwe, M.V., Eds.; Marcel Dekker: New York, NY, 1992; 573-618.
48. Cervone, N. W.; Harper, J. M. *J. Food Proc. Eng.* **1978**, *2*, 83.
49. Harper, J.M. Food Extrusion In *Food Properties and Computer-Aided Engineering of Food Processing Systems*; Singh, R.P.; Medina, A.G., Eds.; Kluwer Academic, 1989; 271-279.
50. Senouci, A.; Smith, A. C. *Rheol. Acta* **1988**, *27*, 546.
51. Vergnes, B.; Villemaire, J. P. *Rheol. Acta* **1987**, *26*, 570.
52. Davidson, V. J.; Paton, D.; Diosady, L. L.; Larocque, G. *Journal of Food Science* **1984**, *49*, 453.
53. Davidson, V. J.; Paton, D.; Diosady, L. L.; Rubin, L. J. *Journal of Food Science* **1984**, *49*, 1154.
54. Jackson, D. S.; Gomez, M. H.; Waniska, R. D.; Rooney, L. W. *Cereal Chemistry* **1990**, *67*, 529.
55. Bueche, F. *Physical Properties of Polymers*; Robert E. Krieger Publishing Company: Huntington, NY, 1979.
56. Rauwendaal, C. *Polymer Extrusion*; Hanser Publishers: Munich, 1986.
57. Cheremisinoff, N. P. *An Introduction to Polymer Rheology and Processing*; CRC Press: Boca Raton, FL 1993.
58. Shogren, R. L. *Carbohydrate Polymers* **1992**, *19*, 83.
59. Karkalas, J.; Raphaelides, S. *Carbohydr. Res.* **1986**, *157*, 215.
60. Evans, I. D. *Starch* **1986**, *38(7)*, 227.

RECEIVED May 24, 1994

Chapter 4

Synthesis and Characterization of Dodecenyl Succinate Derivatives of Saccharides

Homa Assempour[1], M. F. Koenig, and S. J. Huang

Biodegradable Polymer Research Consortium, Institute of Materials Science, University of Connecticut, Storrs, CT 06269–3136

Acid ester derivatives of sucrose, amylose, and corn starch bearing carbon-carbon double bonds were synthesized by reaction of 2-dodecene-1-ylsuccinic anhydride under homogeneous conditions, using a pyridine catalyst. The esters were characterized by FTIR, NMR, elemental analysis, thermal analysis, and solubility tests. The results show that these saccharides can undergo complete or nearly complete substitution in reaction with the anhydride, within the limits of sensitivity of these techniques. The resultant esters posess remarkable solubility in common organic solvents such as THF, methylene chloride, and toluene. Their solubility gradually decreases with time when stored under ambient conditions due to crosslinking through the incorporated double bonds. The esters react wtth aqueous NaOH to give a solution with pH ~8.

A considerable amount of research has been reported in the literature on the development of hydrophobic saccharide esters for a variety of applications (1-5). In the present study, we have synthesized the acid ester derivatives of sucrose, amylose, and corn starch with 2-dodecene-1-ylsuccinic anhydride (DDSA).

$$CH_2CH=CH(CH_2)_8CH_3$$

The purpose of this work was twofold: (1) to increase the hydrophobicity of the saccharides by incorporation of the DDSA with its long alkyl chain. It was hoped that this group would be long enough to exhibit alkyl chain packing, and hence a low

---

[1]Current address: Amir Kabir University of Iran, Tehran, Iran

0097–6156/94/0575–0069$08.00/0

melting transition, ease of processing, and enhanced solubility, and (2) to incorporate reactive functional groups which could provide sites for further processing reactions, such as chain extention, crosslinking, or grafting. Sucrose was chosen as a model compound for the reaction of DDSA with polysaccharides.

Alkenyl succinate, as well as other acid ester derivatives of starch, have been patented (6,7), but little has been published on the synthesis and properties of these derivatives, particularly at a high degree of substitution. This paper focusses on the synthesis of these derivatives.

Experiment

Materials. High-amylose corn starch (HA-CS; 70% amylose) and potato amylose (PA) were obtained from Sigma. The potato amylose was used as received. The HA-CS granules were destructured by heating the starch in DMSO at 100 °C for 1 h. Corn amylose (CA) was isolated by precipitating twice with n-butanol from a 10% solution in DMSO. Sucrose used was pure cane confectioners sugar from Domino. It was dried in a vacuum oven at room temperature prior to use. 2-Dodecen-1-yl succinic anhydride (DDSA; 97% purity) was purchased from Aldrich and used without further purification. Pyridine from Aldrich was distilled and stored over 4 Å molecular sieve until use. All solvents used were of high purity (ACS reagent grade or higher) and used as received.

Esterification of sucrose. A solution of 5 g sucrose in 25 ml DMSO was prepared by heating at 50 °C under an argon stream. To the solution was added 2.5 ml pyridine and 34 g DDSA, and the reaction mixture heated for 16 h at 65 °C. (*Caution: Pyridine is a known mutagen and a very dangerous fire and explosion hazard; standard laboratory procedures should be carefully followed to limit exposure and prevent mishaps.*) The mixture was then poured into 300 ml distilled water and stirred for several minutes to obtain a stable white colloid. On stepwise treatment with an aqueous 0.5 N NaOH solution, a sodium salt was precipitated from the colloid. The precipitant was filtered off, washed with 200 ml acetone, and treated with an aqueous solution of 0.5 N HCl to recover the ester at pH ~7 as a colloid. Final purification was accomplished by extraction of the ester sample with chloroform (*Caution: a CNS poison and carcinogen; avoid inhalation of vapors*). After evaporation of the solvent, the ester was dried in a vacuum oven at 40 °C for one week.

Esterification of amylose. A solution of 5 g CA or PA in 25 or 45 ml DMSO, respectively, was prepared by heating the solution to 110 °C under Ar. The solution was cooled to about 50 °C. Both 2.5 ml pyridine and 24 g DDSA were then added with constant stirring, after which the reaction mixture was heated to 65 °C for 16 h. The mixture was then poured into 300 ml distilled water and shaken. The crude product was separated as a doughy mass from the water by decantation. The product was rinsed in this manner several times with distilled water, after which it was dissolved in 200 ml methanol at 40 °C with stirring. (*Caution: Methanol has a high cumulative toxicity, and can be absorbed through the skin. It is also a potential fire hazard.*) The product was precipitated by cooling in an ice water bath, filtered, and dried in a vacuum oven at room temperature. Final purification was accomplished by twice dissolving in 200 ml acetone, precipitating in distilled water, filtering, and drying in a vacuum oven at 40 °C for one week.

Esterification of HA-CS. One gram of destructured HA-CS, 8 g DDSA, and 10 ml DMSO were combined in a 60 ml test tube. The mixture was then stirred thoroughly with a glass stirring rod. After adding a magnetic stir bar, the test tube was closed with a rubber septum, flushed with argon, and vented with a small needle. The tube was heated in an oil bath to 130 °C and held at that temperature for 6 h. The product was precipitated in 200 ml distilled water. The excess DDSA was suspended in the water by vigorous stirring, causing it to turn milky white. The water was then decanted. The precipitant was rinsed several more times with 50 ml aliquots of distilled water until the water remained clear. The product was then twice dissolved in 100 ml acetone and precipitated in 200 ml distilled water, followed by filtration with a fritted glass funnel. The product was dried in a vacuum oven at 40 °C overnight.

Methods of Characterization. A 270 MHz Bruker NMR spectrometer was used for recording proton decoupled ^{13}C spectra. CH_2Cl_2/DMSO-d_6 was used as solvent, and TMS was added for an internal stardard. The FTIR spectrometer used in these studies was a Nicolet 60 SX. The specimens were prepared by casting thin films on NaCl plates from $CHCl_3$ solutions. Elemental analysis was performed by Galbraith Laboratories. Size-Exclusion Chromatography (SEC) was carried out using a Waters 150C system equipped with two 10^3Å and two 10^5Å Ultrastyragel columns. The mobile phase used was tetrahydrofuran (THF) at a flow rate of 1 ml/min at 30 °C. Differential Scanning Calorimetry (DSC) thermograms were recorded using a Perkin-Elmer DSC-7 at a heating rate of 15 °C/min. A dry nitrogen purge gas was used, and sample sizes of about 10 mg. Thermogravimetric Analysis (TGA) was performed using a Perkin-Elmer TGA-7 at a heating rate of 10 °C/min in a nitrogen atmosphere.

Results and Discussion

Degree of substitution. The extent of substitution of the prepared acid/esters was studied by NMR, elemental analysis, and FTIR. These methods are also useful for detecting impurities in the isolated products. Of special concern was the removal of the pyridine catalyst from the derivatives. Size-exclusion chromatography (SEC) was also performed on the amylose derivatives to determine the extent of degradation of the amylose macromolecules during the esterification reaction.

The ^{13}C NMR results for sucrose, amylose, and their derivatives are shown in Figures 1 and 2. From the results for sucrose, Figure 1, it appears that complete substitution of both the primary and secondary hydroxyl groups has occurred. This is evidenced by the fact that the peaks due to the various carbon atoms of the glucose and fructose subunits, between 60 and 110 ppm, have shifted position. This indicates a change in the chemical environment of the atoms. These peaks also have appeared to have broadened, which may be due to the low signal intensity. This broadening causes some difficulty in comparing the shifts of the individual peaks with the theoretical shifts, because of peak overlap. Peaks from the DDSA component can be clearly seen at 10-40, 120-140, and 170-180 ppm, which are due to the alkyl, alkenyl, and carbonyl carbons, respectively. The splitting of the acid carbonyl peaks at 176.06 and 175.14 ppm and ester carbonyl peaks at 173.97 and 171.92 ppm can be attributed to the incorporation of two environmentally-different acids and esters. This is a consequence of the two different ways that the cyclic anhydride can open, resulting in either the C2 or C5 carbonyl carbon of the anhydride being bonded through oxygen to sucrose. The absence of a peak at 150 ppm indicates that the pyridine catalyst has been removed successfully by the purification procedure.

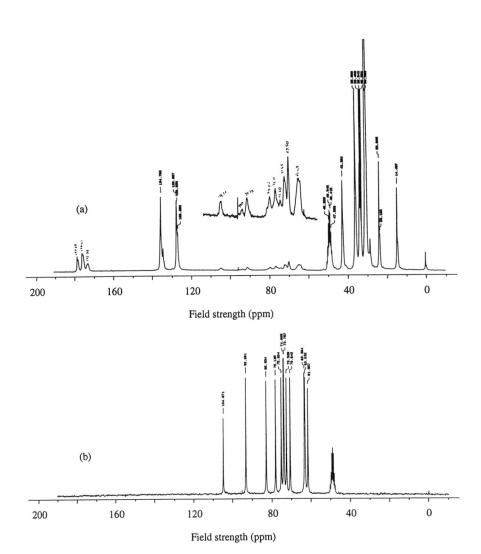

Figure 1. ^{13}C NMR spectra of (a) SU/DDSA and (b) sucrose.

A comparison of the amylose spectra, Figure 2, shows similar features as the sucrose spectra: a shifting of the six amylose carbon peaks, the presence of the DDSA peaks, and the absence of peaks due to pyridine. The peak positions, both measured and calculated, for the amylose carbon peaks are listed in Table I. The shift factors used for these calculations are from ref. 8. Although there are some differences betwen the calculated and measured values, the peak assignments are believed to be accurate. The changes in measured peak positons due to esterification of the C2, C3, and C6 hydroxyl groups show the same trends as those calculated. From these observations, it can be concluded that the PA/DDSA sample has undergone complete substitution, within the detection limits of this technique. The presence of a small C1 peak at 100.8 ppm for the CA/DDSA sample reflects incomplete substitution at the C2 position.

Table II shows the results of the elemental analysis for these derivatives. They are in good agreement with the calculated values and with the NMR results presented above.

The FT-IR spectra of SU/DDSA and PA/DDSA are shown in Figure 3. Both spectra show similar features. Characteristic bands due to carbonyl stretching are visible for both the acid at 1712-1713 cm^{-1} and the ester at 1737-1739 cm^{-1}. The peaks at 1152-1160 cm^{-1} are assigned to C–O stretching of the ester. The absorption bands at 2932 cm^{-1}, 2854 cm^{-1}, and 721 cm^{-1}, due to stretching andbending vibrations of the CH_2 groups, and the peak at 969 cm^{-1}, due to out-of-plane bending of the =C–H bond, are also clearly visible. All of these bands indicate the incorporation of DDSA. The broad band from 3100-3500 cm^{-1} can be attributed to the inter- and intramolecular hydrogen bonding of the carboxylic acid groups.

The SEC data shown in Figure 4 for the two amylose derivatives show that both samples have single peak molecular weight distributions. This implies that the starting materials also have unimodal molecular weight distributions and that the derivatization reaction was homogeneous, with similar conversions for all molecules in the reaction mixtures. It also indicates that not much degradation of the polymer chains occurred during the derivatization reaction, or that any degradation that occurred was uniform throughout the reaction mixture. The calculated molecular weights (M_n = number average, M_w = weight average molecular weights) and polydispersities (M_w/M_n) for these two samples are listed in Table III.

Properties. PA/DDSA and CA/DDSA are opague white solids and SU/DDSA is a light-brown waxy solid. Solubility tests in various ordinary polar and nonpolar solvents showed these acid/esters to be easily dissolved in methylene chloride, chloroform, toluene, THF, and acetone, and partially soluble in ethanol, methanol, hexane, benzene, and carbon tetrachloride. The derivatives are insoluble in distilled water or dilute aqueous acid solutions. On mixing with dilute aqueous NaOH solutions, these acid/esters produce a solution with pH of ~8.

TGA results, Figure 5, show that the derivatives have moderate thermal stability, their onset of decomposition occurring at ~200 °C. DSC thermograms, shown in Figure 6, detect a broad melting transition at about 65 °C. After annealing at 80 °C, a small melting transition was seen at about 90 °C, and a small endotherm at about 112 °C. FT-IR revealed that this endotherm is probably due to the crosslinking reaction of the alkenyl group of the DDSA. This was supported by monitoring the alkenyl absorption band at 969 cm-1 for PA/DDSA while heating at 112 °C. Figure 7 shows the reduction of intensity of this band with time over the course of 12 h.

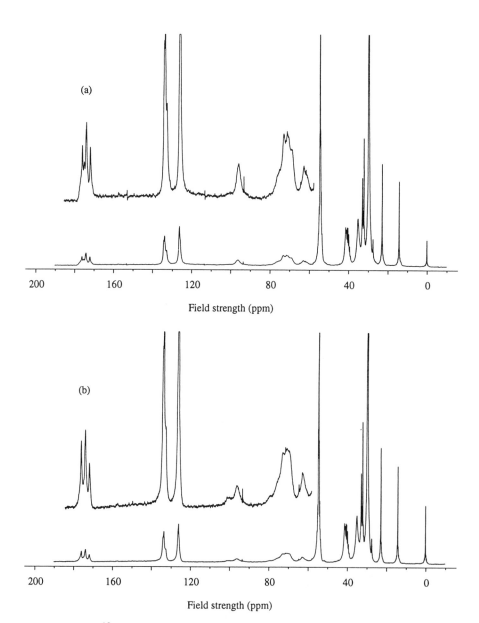

Figure 2. ^{13}C NMR spectra of (a) PA/DDSA, (b) CA/DDSA, and (c) potato amylose.

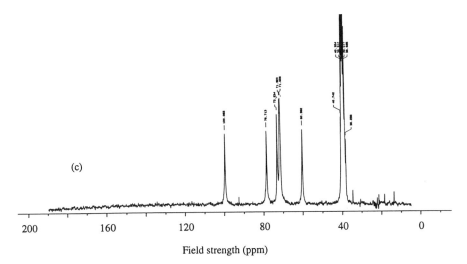

Field strength (ppm)

Figure 2. *Continued*

Table I. Selected ^{13}C NMR chemical shifts of the amylose derivatives

Sample	C1 (ppm)	C4 (ppm)	C5 (ppm)	C2 (ppm)	C3 (ppm)	C6 (ppm)
			Peak Assignments			
PA	100.05	78.71	73.23	71.92	71.60	60.39
(calc.)	121.5	84.7	81.5	77.7	72.9	62.8
PA/DDSA	96.18	75.5	69.7	73.6	71.9	62.96
CA/DDSA	100.8;96.4	75.5	69.7	_a	_a	62.81
ave. shift	-3.78	-3.21	-3.53	+1.68	+0.3	+2.50
(calc.)	-2	-1	-2	+1	+1	+4

a Peak positions uncertain because of peak overlap.

Table II. Composition data obtained from elemental analysis

Sample	Calculated		Experimental	
	C (wt%)	H (wt%)	C (wt%)	H (wt%)
Sucrose	42.10	6.47		
SU/DDSA	67.98	9.37	67.39	9.43
PA	44.44	6.21		
PA/DDSA	67.47	9.15	66.04	9.02
CA/DDSA	67.47	9.15	66.30	9.28

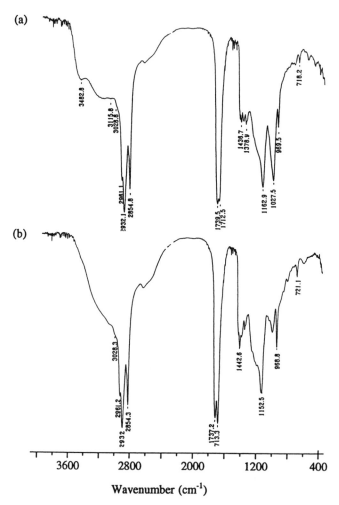

Figure 3. FTIR spectra of (a) PA/DDSA and (b) SU/DDSA.

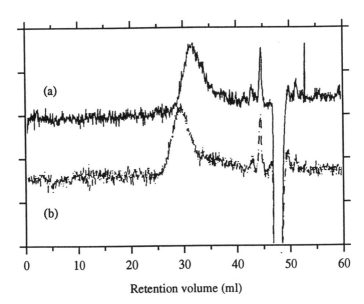

Retention volume (ml)

Figure 4. SEC chromatographs of (a) PA/DDSA and (b) CA/DDSA.

Table III. SEC results for the amylose derivatives[a]

Samples	M_n	M_w	M_w/M_n
PA/DDSA	22,700	70,800	3.11
CA/DDSA	14,300	30,600	2.14

[a] Molecular weights calculated relative to polystyrene standards.

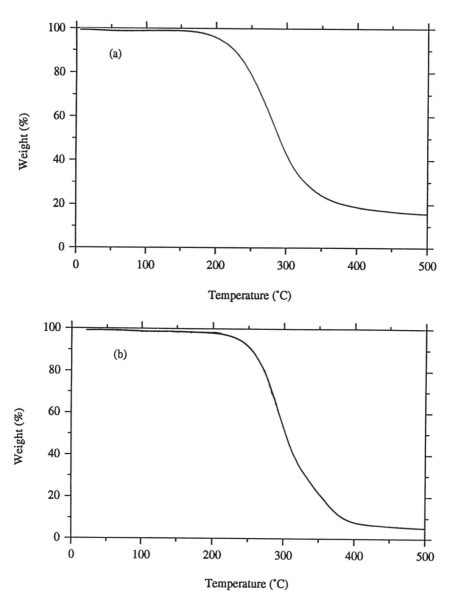

Figure 5. TGA thermograms of (a) PA/DDSA and (b) CA/DDSA.

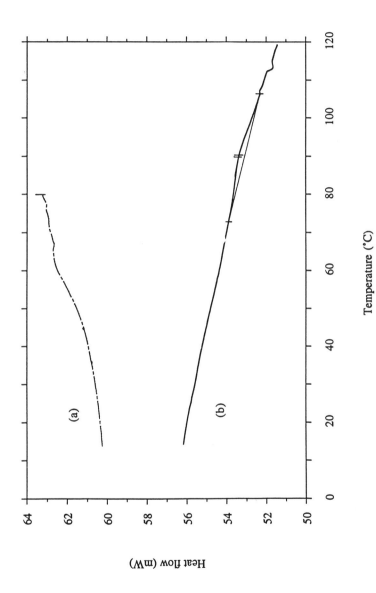

Figure 6. DSC thermograms of PA/DDSA (a) first heating and (b) second heating.

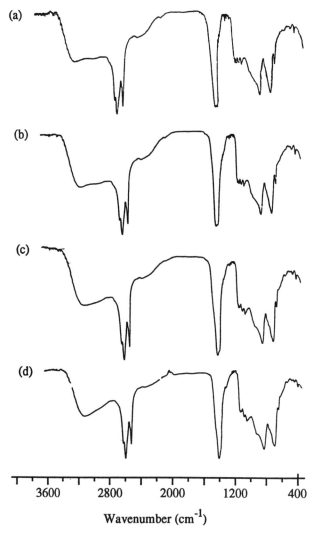

Figure 7. FTIR spectra of PA/DDSA at 112 °C (a) initial spectrum, (b) after 2 h, (c) after 5 h, and (d) after 12 h.

Conclusions

Sucrose, amylose, and starch derivatives of DDSA were successfully made by reacting with an excess of the anhydride, both with and without a pyridine catalyst. It was shown that all of the primary and secondary hydroxy groups in these saccharides are susceptible to substitution, and under the conditions described here, can lead to complete substitution. These derivatives have a high solubility in several common organic solvents. The low melting temperature and reactivity of the alkenyl group makes these materials easy to crosslink at moderate temperatures. The properties of these acid/esters are being further investigated toward their application in designing novel coatings and thermoset polymers, as well as their biodegradability.

Acknowledgments

The authors would like to thank Dr. Levant Cimecioglu for the ^{13}C NMR data and for helpful discussions. One of the authors (H.A.) thanks The Amir Kabir University of Iran for awarding her a Senior Research Fellowship.

Literature Cited

1. *Starch: Chemistry and Technology*,; Whistler, R. L., Bemiller, J. N. & Paschall, E. F., Eds.; ,Academic Press: New York, NY, 1984.
2. *Industrial Polysaccharides: Genetic Engineering,Structure/Property Relations and Applications*; Yalpani, M., Ed.; Elsevier: Amsterdam, 1987.
3. Kawaguchi, T., Nakahara, H. & Fukuda, K. *J. Colloid Interface Sci.*, **1985**, *104*, 290-293.
4. Malm, C. J., Mench, J. W., Kendall, D. L. & Hiatt, G. D. *Ind. Eng. Chem.*, **1951**, *43*, 684-688.
5. Malm, C. J., Mench, J. W., Kendall, D. L. & Hiatt, G. D. *Ind. Eng. Chem.*, **1951**, *43*, 688-691.
6. Caldwell, C. G., U. S. Patent 2 613 206 (1952).
7. Jane, J., Gelina, R. J., Nikolov, Z. & Evangelista, R. L., U. S. Patent 5 059 642 (1991).
8. Silverstein, R. M., Bassler, G. C. & Morrill, T. C. *Spectrometric Identification of Organic Compounds*, John Wiley & Sons: New York, NY, 1981.

RECEIVED July 26, 1994

Chapter 5

Mechanical Properties of Pectin–Starch Films

D. R. Coffin and M. L. Fishman

Eastern Regional Research Center, Agricultural Research Service,
U.S. Department of Agriculture, 600 East Mermaid Lane,
Philadelphia, PA 19118

Films were made from plasticized blends of citrus pectin and high amylose starch, and characterized by dynamic mechanical analysis to determine the effect of composition on film mechanical properties. The films were cast from water onto glass plates, allowed to dry, and removed. Mechanical analysis was done using a Rheometrics RSA II solids analyzer. The blends formed excellent films whose mechanical properties were highly dependent on both plasticizer level and pectin/starch ratio. The behavior was explained on the basis of pectin-starch and pectin-plasticizer interactions.

Biopolymers are increasingly being studied and used for applications where synthetic polymers have traditionally been the materials of choice. The heightened interest in these materials comes from the desire to increase the use of biodegradable and recyclable materials in order to limit the amount of material sent to landfills, as well as to use renewable raw material resources instead of non-renewable petroleum sources. One promising system of this type consists of plasticized blends of high methoxyl pectin and high amylose starch.

The use of pectin and starch for free standing films, and possibly other fabricated shapes, represents a potential economic benefit to farmers and agricultural processors. The successful commercialization of products using them would represent new markets for agricultural products, as well as for waste by-products which currently have no commercial value and must be disposed of at additional cost.

Low methoxyl pectin films were first made and characterized in the 1940's (*1-3*). They required the use of calcium or other multivalent cations as cross-linking agents. These films exhibited fair mechanical properties, but had poor folding endurance, and little subsequent work was done with them. Also, pectins have been investigated as coatings for fruits and nuts (*4, 5*).

Starch films have been investigated since the 1950's (*6-8*). These have shown interesting mechanical properties, however, they are moisture sensitive and tend to embrittle with time. More recent work by Otey and others (*9-11*) has shown that the addition of thermoplastics at fairly high levels results in films that are highly flexible and do not embrittle. High amylose starches are expected to give better films than typical amylose level starches (*12*).

Previous work in our laboratory has shown that films made from high methoxyl lime pectin and high amylose starch have very good mechanical properties and appear to be suitable for use in applications where strong biodegradable films are advantageous (*13*). In that study we showed that the best pectin for use in films has the highest molecular weight and degree of methyl esterification. The use of glycerine or other plasticizer is necessary to make a sufficiently flexible and non-brittle film. Addition of a large proportion of high amylose starch resulted in only modest decreases in the dynamic mechanical properties (storage modulus and loss modulus) of the films, had a beneficial effect on their surface properties, and reduced material costs. Films containing only starch and glycerine are extremely brittle.

The present work expands on our previous studies and more fully characterizes the effect of composition on pectin/starch film properties. Citrus pectin of sufficiently high methoxyl content and molecular weight is shown to have mechanical properties essentially equivalent to those of the high methoxyl lime pectin films studied earlier. The lime pectin is no longer commercially available, so it was decided to use the most similar commercial citrus pectin. Furthermore, a wide range of film properties was obtained by the use of a larger range of plasticizer levels. In addition, increased emphasis was placed on the determination of low temperature mechanical properties and thermal transitions.

Experimental

Materials. MexPec 1400, a citrus pectin with a degree of methyl esterification of 71%, was provided by Grindsted Products, Inc. (Kansas City, KS) and was used as received. It is identified as DM71.

Amylomaize VII (ca. 70% amylose, 30% amylopectin) was provided by American Maize-Products Co. (Hammond, IN). It was used as received.

Glycerine was ACS reagent grade purchased from Aldrich Chemical Co. (Milwaukee, WI).

Water was HPLC grade prepared using a Modulab Polisher I water system (Continental Water Systems, Inc.).

Film Preparation. Films were prepared by mixing solutions of pectin and glycerine with gelatinized starch solutions, casting them on a glass plate using a "Microm" film applicator (Paul N. Gardner Co., Pompano Beach, FL) and allowing the films to air dry overnight. After air drying the samples were vacuum dried for 30 min at room temperature to remove residual water. Films were removed from the plates with a razor blade. Wet film thicknesses of 2-2.5 mm were used, giving dry film thicknesses of 0.04-0.05 mm.

The pectin was dissolved by slowly adding a measured amount of it to 20 or 25 ml of HPLC grade water with stirring. The glycerine was added to the water prior to the addition of pectin. The solutions were stirred for one to two hours until all of the pectin appeared to be dissolved. The total concentration of pectin and glycerine in the solutions was in the range of 5 to 7% by weight depending on the formulation.

Gelatinized starch solutions were prepared by mixing the appropriate amount of starch (0.05 to 0.67 gm) with 10 ml of HPLC grade water in a Parr microwave bomb (Parr Instrument Co., Moline, IL), and heating in a 700 watt Amana Model R321T Radarange microwave oven for three minutes at 50% of full power.

The gelatinized starch solutions were cooled in a water bath at room temperature for 25 min and then added, still warm, to the pectin solutions with stirring. No significant amount of retrogradation of the starch was observed during cooling as judged by the relative clarity of the solutions. The mixtures were allowed to stir for an additional hour prior to casting.

Mechanical Testing. Mechanical testing was done on a Rheometrics RSA II Solids Analyzer (Piscataway, NJ) using a film testing fixture. Testing was done within one day of sample preparation. Both tensile tests and dynamic mechanical analysis were carried out. Test samples were cut from the films with a razor blade. Nominal sample dimensions were 6.4 mm x 38.1 mm x 0.04 mm. Sample thickness was measured with a micrometer and sample width was measured with a millimeter ruler. The gap between the jaws at the start of each test was 23.0 mm. Data analysis was carried out using the Rheometrics RHIOS software. At least three test runs were made for each sample. The variability between replicates for the tensile measurements was approximately 25%. Variability between replicates in the dynamic measurements was about 10%. The curves shown in the Figures are representative of those obtained.

Tensile data were obtained at room temperature using a strain rate of 0.005 sec^{-1} (30%/min). Tensile strength, initial modulus, and elongation to break were measured. A maximum elongation of 13% could be obtained due to instrument constraints.

For dynamic mechanical measurements a nominal strain of 0.1% was used in all cases, with an applied frequency of 10 rad/sec (1.59 Hz). Storage modulus (E'), loss modulus (E"), and loss tangent (tan δ) were determined as a function of temperature. Data were taken over the range of -100°C to 200°C using a heating rate of 10°C/min. The samples were equilibrated under dry nitrogen at the starting temperature prior to running the test. This, along with the vacuum drying of the films, was felt to minimize the amount of moisture present.

Results and Discussion

Tensile data (tensile strength, initial modulus, and elongation to break) were obtained on samples with a pectin/starch ratio of 90/10 and glycerine levels of 16 to 75%. Typical tensile curves are shown in Figure 1 for each composition.

For the two lowest glycerine levels, 16 and 30%, elongation to break was less than 2% and the tensile strength was about 2.5×10^8 dynes/cm^2. With 45% glycerine the elongation increased to about 5% and the tensile strength to 3.5×10^8 dynes/cm^2. A much greater change occurred when the glycerine level was increased to 60%. Elongation increased to more than 13% (the instrument limit) and the tensile strength decreased to less than 2×10^8 dynes/cm^2. At 75% glycerine the tensile strength dropped to 3×10^7 dynes/cm^2. The initial modulus of the sample with 16% glycerine was 3.4×10^{10} dynes/cm^2. There was a linear decrease in modulus with increasing glycerine level up to 60% glycerine. Increasing the glycerine level to 75% further lowered the modulus by an order of magnitude.

In dynamic mechanical analysis the storage modulus (E') is defined as the stress in phase with the strain divided by the strain, and is a measure of the energy stored and recovered per cycle. The loss modulus (E") is defined as the stress 90° out of phase with the strain divided by the strain, and is a measure of the energy dissipated per cycle of the deformation *(14)*.

Figure 2 shows the effect of glycerine level on the storage modulus and loss modulus of films with a pectin/starch ratio of 90/10. The storage moduli of the samples were all essentially the same at -100°C. As the temperature increased the values for the moduli decreased, with the reduction being greater as the glycerine level was raised. There was a tendency for the modulus to plateau near room temperature, followed by a further decrease at higher temperatures. This plateau was most noticeable at the lower glycerine levels. The samples containing 60 and 75% glycerine could not be tested past 100-125°C because of their very low strengths at these temperatures. The dip seen near 100°C in some of the samples with higher glycerine levels may have been related to the presence of some water.

A nearly identical trend was seen for the loss modulus. In addition, there was a peak which shifted from -55° at 16% glycerine to -75° at 75% glycerine. The area under the peak was proportional to the glycerine concentration in the films, and indicates a transition involving or caused by the glycerine.

When the pectin/starch ratio was lowered to 70/30 there was a noticeable change in the effect of glycerine on the film properties. These data are shown in Figure 3. At temperatures up to about 0°C the effect of glycerine was very similar to that seen with the 90/10 pectin starch ratio. However, above this temperature the moduli for the samples containing 60 and 75% glycerine did not decrease nearly as much as expected. The glycerine related peak in the loss

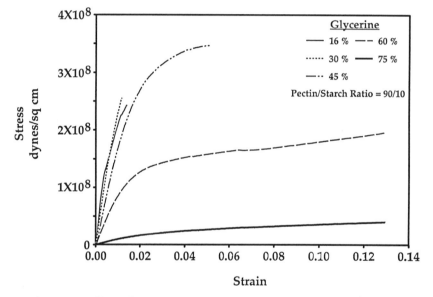

Figure 1. Effect of glycerine level on tensile properties for 90/10 blends of pectin DM71 and Amylomaize VII.

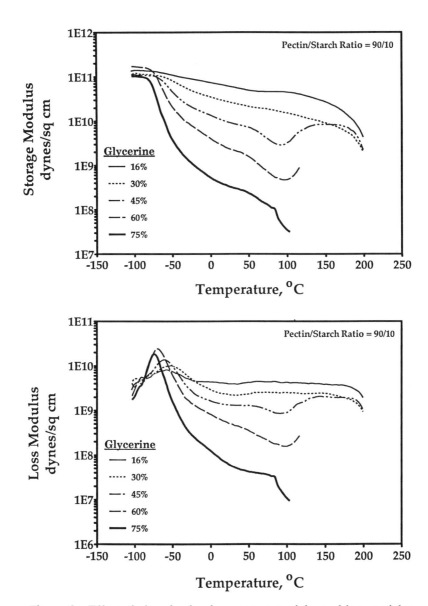

Figure 2. Effect of glycerine level on storage modulus and loss modulus for 90/10 blends of pectin DM71 and Amylomaize VII.

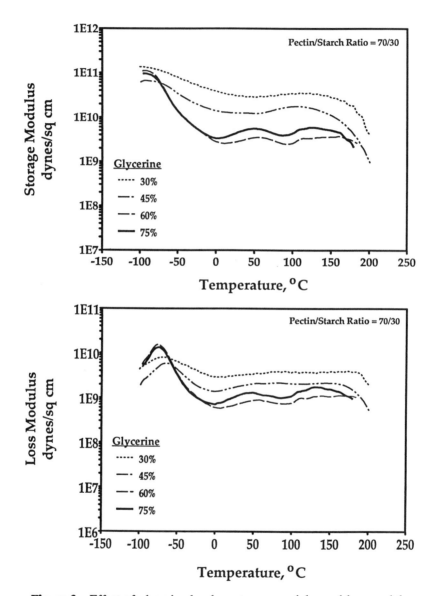

Figure 3. Effect of glycerine level on storage modulus and loss modulus for 70/30 blends of pectin DM71 and Amylomaize VII.

modulus curves showed a similar concentration dependence to that seen with the lower starch level.

At the 55/45 pectin/starch ratio quite different behavior was observed as indicated by the data in Figure 4. There was essentially no effect of glycerine level on either the storage or loss modulus, and the temperature dependence of the peak maximum in the loss modulus curve was diminished.

In all of the samples, the drop off in storage and loss modulus above 180°C appeared to have two causes, sample softening and degradation of the biopolymers. When the sample chamber was opened prior to a sample attaining 180°C there was little or no discoloration of the material, however samples taken to 190-200°C were found to be dark brown, indicating degradation.

Electron microscopy has revealed that pectin in pure water exists as a cross-linked gel held together by hydrogen bonds (*15,16*). In 50% aqueous glycerine this gel is dissociated into its component parts which are for the most part rods, segmented rods, and kinked rods. Image analysis of electron micrographs of pectin before and after dissociation reveals that this gel is cross-linked at the rhamnose inserts in the pectin backbone (*16*). Analysis of high amylose starch by high performance size exclusion chromatography with online viscometric detection indicates that our process of gelatinization produces dense spherical starch structures (*17*). At high pectin/starch ratios and about 45% glycerine, the effect of glycerine on the films changes as indicated by increasing elongation to break and decreasing moduli.

Evidence from microscopy and chromatography suggests that below 45% glycerine, pectin/starch films are essentially pectin gels partially filled with dense starch particles, whereas above 45% glycerine, starch interacts with dissociated pectin rods, segmented rods, and kinked rods. Possibly, preferential interactions between pectin and glycerine produce glycerine plasticization of pectin chains allowing films to elongate by pectin chain slippage when they are stressed. Decreasing the ratio of pectin to starch may tend to stabilize pectin gels by increasing starch-pectin interactions at the expense of glycerine-pectin interactions. Thus as the pectin/starch ratio decreases at constant glycerine, the moduli will tend to increase and elongation to break will decrease.

Conclusions

Plasticized blends of citrus pectin and high amylose starch make strong flexible films with properties highly dependent on composition. The tensile strength can be varied by an order of magnitude, while static modulus and dynamic modulus can be varied by more than two orders of magnitude. In addition, the elongation to break increases as the moduli decrease. At low starch levels increasing the glycerine level dramatically lowers the moduli, but at higher starch levels the effect of glycerine is reduced. The films are thermally stable up to 180°C.

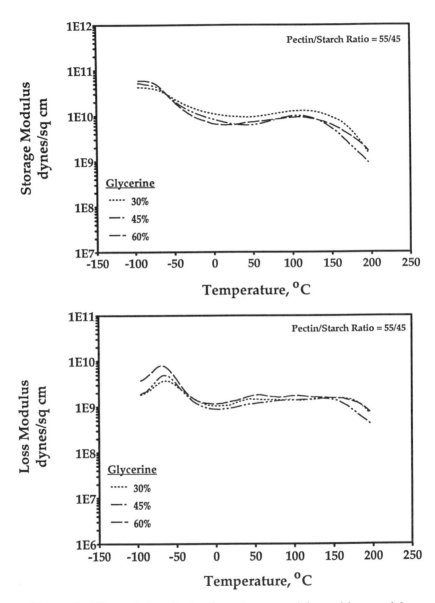

Figure 4. Effect of glycerine level on storage modulus and loss modulus for 55/45 blends of pectin DM71 and Amylomaize VII.

Literature Cited

1. Maclay, W. D.; Owens, H. S. *Modern Packaging*, **1948**, *22*, 157-158.
2. Schultz, T. H.; Owens, H. S.; Maclay, W.D. *J. Colloid Sci.*, **1948**, *3*, 53-62.
3. Schultz, T. H.; Miers, J. C.; Owens, H. S.; Maclay, W. D. *J. Phys. Colloid Chem.*, **1949**, *53*, 1320-1330.
4. Miers, J. C.; Swenson, H. A.; Schultz, T. H.; Owens, H. S. *Food Technol.*, **1953**, *7*, 229-231.
5. Swenson, H. A.; Miers, J. C.; Schultz, T. H.; Owens, H. S. *Food Technol.*, **1953**, *7*, 232-235.
6. Wolff, I. A.; Davis, H. A.; Cluskey, J. E.; Gundrum, L. J.; Rist, C. E. *Ind. Eng. Chem.*, **1951**, *43*, 915-919.
7. Rankin, J. C.; Wolff, I. A.; Davis, H. A.; Rist, C. E. *Ind. Eng. Chem. Chem. Eng. Data Ser.*, **1958**, *3*, 120-123.
8. Mark, M. A.; Roth, W. B.; Mehltretter, C. L.; Rist, C. E. *Food Technol.*, **1966**, *20*, 75-77.
9. Otey, F. H.; Westhoff, R. P.; Russell, C. R. *Ind. Eng. Chem. Prod. Res. Dev.*, **1977**, *16*, 305-308.
10. Otey, F. H.; Westhoff, R. P.; Doane, W. M. *Ind. Eng. Chem. Prod. Res. Dev.*, *19*, **1980**, 592-595.
11. Otey, F. H.; Westhoff, R. P.; Doane, W. M. *Ind. Eng. Chem. Res*, **1987**, *26*, 1659-1663.
12. Young, A. H.   In *Starch Chemistry and Technology; 2nd* ed., Whistler, R. L.; BeMiller, J. N.; Paschall, E. F., Eds. Academic Press, New York, NY, 1984, p. 269.
13. Coffin, D. R.; Fishman, M. L. *J. Agric. Food Chem.*, **1993**, *41*, 1192-1197.
14. Ferry, J. D. *Viscoelastic Properties of Polymers*, John Wiley & Sons, New York, NY, 1961, pp. 30-31.
15. Fishman, M. L.; Cooke, P.; Levaj, B.; Gillespie, D. T.; Sondey, S. M.; Scorza, R. *Arch. Biochem. Biophys.*, **1992**, *294*, 253-260.
16. Fishman, M. L.; Cooke, P.; Hotchkiss, A.; Damert, W. *Carbohydr. Res.*, **1993**, *248*, 303-316.
17. Fishman, M. L.; Hoagland, P. D. *Carbohydr. Polym.*, **1994**, in press.

RECEIVED May 21, 1994

Chapter 6

Biodegradable Plastics Made from Agricultural Biopolymers

J. Jane, S. Lim, I. Paetau, K. Spence, and S. Wang

Center for Crops Utilization Research and Department of Food Science and Human Nutrition, Iowa State University, Ames, IA 50011

Biopolymers (e.g., starch, protein, and cellulose) from various agricultural sources have been investigated for manufacturing biodegradable plastics. Cross-linking of starch and zein mixtures and of soy proteins have shown that the tensile strength and water resistance of the plastics can be increased. Plastics made from dialdehyde starch and zein mixtures displayed particularly improved water resistance. Cereal flours are economically competitive materials for the plastics. With proper processing, both soy isolate and soy concentrate displayed positive properties for extrusion and injection molding of the plastics. Cellulose fibers can be used as extenders in the plastics to improve tensile strength and water resistance.

Starch, protein, and cellulose are all biopolymers. Cellulose, the most abundant biomass, has been extensively used by the paper and other industries. Starch is the second-largest biomass produced on earth. Starch consists of α 1-4 linked d-glucopyranose units and has a molecular weight of up to 10^7 daltons (1). With the configuration of α 1-4 linked glucopyranoses, starch displays a random-coil conformation. The molecules consist of hydroxyl groups at the equatorial positions and hydrogen at the axial positions of the glucopyranose ring. The random-coiled starch molecule tends to form helices by interacting with hydrophobic molecules to reduce the exposure of its hydrocarbon moiety to surrounding water molecules. Two types of helices are commonly observed: one is the double helix formed between two starch molecules (retrogradation) and the other is single helix formed by complexing with other chemicals carrying hydrophobic groups, such as ethanol and butanol. At temperatures above the glass-transition temperature (T_g), starch molecules have a tendency to associate and crystallize (retrograde) (2,3). The retrogradation rate increases when the starch concentration increases and the temperature decreases to between 0 and 5°C (1).

0097–6156/94/0575–0092$08.00/0
© 1994 American Chemical Society

Proteins, such as soy protein and zein (a hydrophobic corn protein), are polymers of various amino acids. Different amino acid compositions render different properties to materials in which they are used. Soy protein contains a significant amount (~35%) of acidic amino acids (glutamic acid, or glutamine, and aspartic acid, or asparagine) and basic amino acids (lysine and arginine) (4) and is soluble in salt and in alkaline solutions. Around its isoelectric point (pH 4.5), soy protein carries minimal charges and is insoluble in water. Zein contains high concentrations of hydrophobic amino acids (leucine, proline, and alanine, ~35%) as well as glutamic acid (24%) (5) and, therefore, is soluble in 70% alcohol instead of in water.

Because of its high concentration of hydroxy groups and its tendency to retrograde at temperature $\geq T_g$, starch alone does not produce desirable plastics. Plastics made from starch alone are water sensitive, readily disintegrate in water, and lack storage stability. To overcome these problems, starch has been mixed with other synthetic polymers (6) or with biopolymers (7,8) for plastic materials.

Plastics made exclusively from biopolymers are suitable for making disposable products, such as food service utensils, containers, and outdoor sporting goods. Once used, the disposable utensils and containers can be collected, processed, and reused for animal feed or as soil conditioners. Left in the field, outdoor sporting goods, such as golf tees, will degrade within a short period of time, and the amino acids released from the material will fertilize or condition the soil.

Biodegradation studies of the plastic materials made from mixtures of starch and protein by respirometry have shown that the plastic materials were degraded faster than their parent materials, starch or protein, alone (9). This can be attributed to a balanced nutrient mixture of starch (carbon source) and protein (nitrogen source) in the material and to the molding process. The molding process results in denatured protein and gelatinized starch, which are more susceptible to microbial degradation.

In this paper, we intended to provide an overview of what had been explored in our laboratory on the development of biodegradable plastics from agricultural-based biopolymers. The plastics in this study were made from cross-linked starch and zein mixtures, cereal flours, dialdehyde starch and protein mixtures, soy isolate (containing more than 90% protein), soy concentrate (containing more than 70% protein), and soy protein and cellulose mixtures.

Biodegradable Plastics Made from Cross-Linked Starch and Zein Mixtures

Mixtures of starch and various proteins have been tested for plastic processing. An injection molded plastic made from corn starch and zein mixture displays high water sensitivity and disintegrates in water within a few hours (7).

Starch and protein mixtures cross-linked by adipic/acetic anhydrides, by formaldehyde, and by glutaraldehyde produced plastics with enhanced tensile strength and percentage of elongation and reduced water absorption (8). Properties of glutaraldehyde cross-linked corn starch/zein plastics are shown in Table I.

Biodegradable Plastics Made from Cereal Flours

Cereal flours, containing different proportions of starch, protein, and lipids, can be used for manufacturing biodegradable plastics. Experimental results showed that certain solvent-soluble compounds (e.g., phenolic compounds and unsaturated fatty acids) triggered oxidation and degradation reactions at elevated temperatures during processing. Removal of these substances significantly improved tensile properties and the color of the molded plastics. Examples are given in Table II. Cross-linking of the solvent-treated flour further increased tensile strength (Table II).

Table I. Effect of Glutaraldehyde Concentration on Properties of
Corn Starch-Zein (9:1) Plastics[a]

Glutar-aldehyde[b] (%)	Tensile Strength[c] (kg/mm^2)	Percent Elongation[c]	% Water Absorption	
			2 hrs	24 hrs
0	0.6(0.1)	1.2(0.2)	75.0(0.5)	disintegrate
0.25	1.2(0.1)	1.2(0.2)	10.1(0.4)	disintegrate
0.5	1.7(0.6)	1.8(0.3)	7.7(0.2)	32.3(2.5)
1.0	1.6(0.2)	1.6(0.4)	7.0(0.3)	22.6(0.6)
1.5	2.2(0.5)	1.8(0.6)	6.6(0.1)	21.0(0.4)
2.0	2.5(0.3)	2.0(0.4)	7.9(0.1)	26.1(0.2)

[a] Numbers are averages of the test data of five ASTM standard articles. ( ) = standard deviation. 75% aqueous methanol (1:1.3 = 75% methanol:total dry weight of the solid mixture) was used to dissolve zein.
[b] Percentage weight of glutaraldehyde is calculated on the basis of total dry weight of zein and corn starch.
[c] Tensile strength and percent elongation were obtained at the breakage of the specimen.

Biodegradable Plastics Made from Dialdehyde Starch and Protein Mixtures

Dialdehyde starch can be prepared by treating native starch with sodium meta-periodate solution (0.11 M) (10). Periodate ions break bonds between C2 and C3 of glucopyranose rings and generate two aldehyde groups. When mixed with proteins and processed at high temperature and pressure, the dialdehyde starch cross-links with protein. The aldehyde groups can react with hydroxyl groups to form hemiacetals and acetals, can react with amino groups to form schiff bases, and also can react with sulfhydryl groups.

Plastics made from mixtures of dialdehyde starch and zein displayed increased tensile strength and percentage of elongation (Table III). The plastics also displayed excellent water resistance (Table III). After 24 hr submersion in water at 25°C, a plastic made from 25% zein and 75% dialdehyde starch absorbed ~5% water. The color of the material darkened after molding, indicating that oxidation reactions took place during heat treatment. With the presence of antioxidants, both color and physical strength were improved (Table III).

Table II. Tensile Properties and Water Absorption of the Compression-Molded Native, Solvent-Treated, and Cross-Linked Flours

Cereal Flours	Solvents	TS[a] (Kg/mm²)	Elongat. (%)	Young's Modules (kg/mm²)	% Water Absorp.[b] 2 hrs	24 hrs
Corn	None	0.2(0.0)[c]	0.6(0.1)	103(19)	Disint.[d]	Disint.
	Ethanol	0.9(0.2)	0.9(0.1)	153(74)	Disint.	Disint.
	n-Hexane	1.7(0.4)	1.3(0.3)	211(10)	Disint.	Disint.
	Chloroform	1.8(0.1)	1.4(0.5)	201(7)	Disint.	Disint.
Wheat	None	0.7(0.2)	0.9(0.1)	90(9)	12.1(1.3)	Disint.
	Ethanol	1.9(0.3)	2.3(0.2)	181(9)	12.8(0.2)	42.5(0.9)
	n-Hexane	1.5(0.3)	1.6(0.1)	195(11)	12.5(0.4)	41.5(1.7)
	Chloroform	1.6(0.0)	1.9(0.3)	186(13)	12.7(0.3)	41.9(0.6)
Wheat	Ethanol & Cross-link[e]	2.7(0.6)	2.9(0.3)	326(68)	24.2(1.6)	62.8(0.7)

[a] Tensile strength at breakage (TS), percent elongation at breakage, and Young's modulus were measured according to ASTM D 638-86 procedure.
[b] Percent water absorption was measured according to ASTM D 570-81 procedure.
[c] ( ) = standard deviation.
[d] Disintegrated.
[e] Cross-linked with adipic/acetic anhydride.

Biodegradable Plastics Made from Soy Isolate

Soy isolate consists of more than 90% protein and has proven thermoset properties. Henry Ford, back in the 1930s and 1940s, mixed soy protein with phenol-formaldehyde resin to produce automobile body parts (11). Brother and McKinney reported making plastics by using soy protein and various cross-linking agents (12). Plastics made from molding of soy protein display swelling after submersion in water. To make plastics of improved tensile strength, water resistance, and other physical properties, we investigated the following processing parameters:

Cross-Linking Effect. Cross-linking reactions applied to soy protein improved the tensile properties. The tensile strength of the soy-protein-based plastics increased and water absorption decreased as formaldehyde concentration increased (Table IV). The percentage of elongation decreased at low formaldehyde concentrations (1 and 2.5%) but increased at high concentrations (5%). Beyond 1% of formaldehyde concentration,

Young's modulus plateaued. These results were in agreement with those reported by Brother and McKinney (12).

Treating soy protein with glyoxal, a cross-linking agent, did not increase tensile strength, but it reduced the percentage of elongation as well as water absorption (13). The color of the cross-linked material darkened during molding, indicating that oxidation reactions took place and resulted in decomposition during molding.

Table III. Effect of Starch Oxidation (Dialdehyde Starch) on Properties of the Plastics[a]

Starch Oxidation (%)	Tensile Strength (kg/mm²)	Elongation (%)	Young's Modulus (kg/mm²)	Water Absorpt. (%, 24h)
0	1.9 (.2)[b]	2.3 (.2)	130 (7)	42.6 (2.3)
1	2.1 (.1)	2.5 (.1)	142 (6)	26.9 (1.1)
5	2.2 (.2)	2.8 (.1)	157 (7)	26.5 (.3)
9	2.0 (.3)	2.0 (.3)	159 (5)	29.7 (1.1)
90	3.9 (.4)	4.0 (.5)	180 (11)	5.3 (.3)

[a] The plastic specimens were made from 25% zein and 75% starch dialdehyde. Numbers are averages of the test data of five ASTM standard articles.
[b] ( ) = standard deviation.

PH Effect. At pH 4.5, near the isoelectric point of soy protein, the protein is known to have the least water solubility. On this basis, soy protein pH was adjusted to its isoelectric point before molding, and water absorption of the protein plastic significantly decreased. Various acids (i.e., hydrochloric acid, sulfuric acid, propionic acid, and acetic acid) were used for pH adjustments. Results showed that plastic tensile strength slightly decreased after sulfuric acid and hydrochloric acid treatments, whereas acetic and propionic acids caused no loss of tensile strength (Fig. 1). Propionic acid displayed the best performance on tensile properties of the molded specimens among all the acids tested. All the acid treatments (pH 4.5) resulted in similar decreases in water absorption (Fig. 1), which suggested that an isoelectric effect was the major factor in improving water resistance.

Temperature Effect. Processing temperature is critical for tensile properties and the water absorption of plastics made from soy protein. Biopolymers, in general, have relatively low decomposition temperatures. Thus, if the processing temperature exceeds the decomposition temperature, tensile strength and elongation will decrease, and the processed material will more readily absorb water. Examples of tensile properties affected by the processing temperatures of soy protein (10% moisture content) are shown in Fig. 2. The results showed that 160°C molding produced the greatest tensile strength (stress) and elongation (strain). At 180°C, decomposition might have taken place and both stress and strain decreased. Water absorption also increased after processing at 180°C.

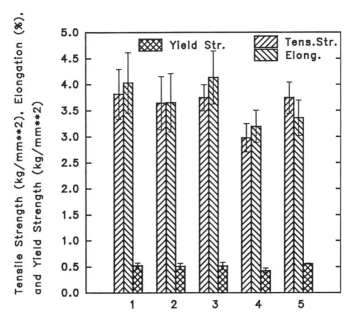

Figure 1. Properties of soy isolate plastic affected by acid treatments (1. control; 2. HCl; 3. propionic acid; 4. H_2SO_4; and 5. acetic acid. 8% moisture content).

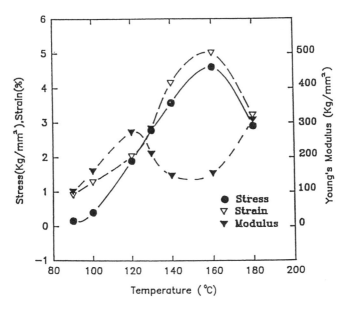

Figure 2. Temperature effect on tensile properties of soy isolate plastics.

Table IV. Effect of Formaldehyde Concentration on Properties of
Soy Isolate Plastics[a,b]

% Form- aldehyde	Tensile Strength (kg/mm^2)	Elongation (%)	Yield Strength (kg/mm^2)	Young's Modulus (kg/mm^2)	Water Absorp. (%, 26h)
0	3.7 (.2)	4.2 (.5)	.52 (.02)	144 (8)	31.7 (.4)
1.0	4.1 (.1)	3.5 (.1)	.54 (.03)	173 (8)	29.4 (.4)
2.5	4.0 (.2)	3.6 (.2)	.56 (.02)	174 (6)	26.1 (.9)
5.0	4.9 (.2)	4.4 (.6)	.68 (.01)	175 (9)	23.2 (.6)

[a] Numbers are averages of the test data of five ASTM standard articles. ( ) =
standard deviation.
[b] The specimens were molded with 8% moisture content at 160°C.

Moisture Content Effect. The optimal molding temperature varied with the moisture
content of the soy protein material. Molding temperature increased as the moisture
content in the raw material decreased. Moisture content in the protein also affected
physical properties of the molded plastic, including tensile strength, elongation,
Young's modulus, and water absorption.   Moisture is required for molding
biopolymers, however, too much moisture present in the molding material resulted in
meta-stable plastic products with inferior physical properties (13). Shapes of the
products also changed with time as moisture was lost through evaporation.

Water absorption of molded plastics decreased as the original moisture content
of the molding material mixture increased. Scanning electron micrographs showed
that the plastics molded with greater moisture contents displayed smoother surfaces
than those made with lower moisture content (13). The results suggested that the
protein polymer molecules were better aligned at a higher moisture content, perhaps
through charge and hydrophobic interactions. The molded plastic absorbed less
moisture when immersed in water.

Biodegradable Plastics Made from Soy Concentrate

Soy concentrate consists of more than 70% protein with the remainder being complex
polysaccharides. Plastics made from soy concentrate displayed higher tensile strength
than those made from soy isolate. The water absorption of soy-concentrate plastics,
however, also was higher than that of soy-isolate plastics (Table V). This could be
attributed to the complex polysaccharides present in soy concentrate.   The
carbohydrate may serve as a binder to increase the strength. The hydrophilic nature
of the complex carbohydrates, however, increase the water absorption of the plastics.

Table V. Effect of Molding Temperature on Properties of Plastic
Specimens Made from Soy Concentrate[a,b]

Molding Temp. (°C)	Tensile Strength (kg/mm²)	Elongation (%)	Yield Strength (kg/mm²)	Young's Modulus (kg/mm²)	Water Absorp. (%, 26h)
100	3.0 (.3)	3.1 (.3)	.46 (.06)	166 (5)	137 (4)
120	3.7 (.5)	3.7 (.3)	.53 (.08)	163 (6)	119 (3)
140	4.1 (.6)	4.0 (.5)	.58 (.08)	166 (3)	98 (2)
160	4.1 (.2)	3.9 (.2)	.55 (.02)	167 (12)	89 (2)

[a] Numbers are averages of the test data of five ASTM standard articles. ( ) = standard deviation.
[b] The specimens were molded with a moisture content of 11.2%.

Biodegradable Plastics Made from Soy Protein and Cellulose

Microcrystalline (ca. 0.4µm long), short (ca. 1mm long), and long (ca. 3.5mm long) fibers were incorporated into soy isolate for plastic manufacturing. Results of compression-molded plastics showed that short fiber increased tensile strength and decreased water absorption, whereas microcrystalline and long fibers decreased tensile strength and increased water absorption (9).

Summary

Biopolymers, including starch, protein, and cellulose, have demonstrated great potential for manufacturing biodegradable plastics. Mixtures of these polymers with different combinations and at different proportions have produced plastics with improved physical properties and at reduced production costs.

Literature Cited

1. French, D. in MTP International Review of Science: Biochemistry of Carbohydrates, Biochemistry Series One, Vol.5, Ed. W. J. Whelan, Butter-Worths, London, pp 269-335.
2. Jane, J. ; Robyt, J. F. Carbohydr. Res. 1984, 132, 105-118.
3. French, A. D.; Murphy, V. G. Cereal Foods World, 1977, 22, 61.
4. Nielsen, N. C. In New Protein Foods; Altschul, A. M.; Wilcke, H. L. Eds.; Academic Press: Orlando, FL, 1985, Vol. 5; pp 27-64.
5. Paulis, J. W. Cereal Chem. 1981, 58, 542-546.
6. Lay, G.; Rehm, J.; Stepto, R.F.; Thoma, M.; Sachetto, J.-P.; Lentz, D. J.; Silbiger, J.; U. S. Patent 5,095,054; 1992.
7. Cole, E. T.; Daumesnil, R.; UK Patent 2,214,920A, 1989.

8.   Lim, S.; Jane, J. In Carbohydrates and Carbohydrate Polymers; Yalpani, M. Ed.;
     ATL Press: Mount Prospect, IL, 1993, pp 288-297.
9.   Spence, K. E., M. S. Thesis "Biodegradable plastics made from dialdehyde starch
     and zein." Iowa State University, Ames, Iowa. 1994.
10. Mehltreeter, C. L. In Methods in Carbohydrate Chemistry; Whistler, R. L. Ed.;
     Academic Press: Orlando, FL, 1964, Vol. 4; pp 316-317.
11. Anonymous, Time, p.63. August 25, 1941.
12. Brother, G.H.; McKinney, L.L. Ind. Eng. Chem. 1940, 32, 1002-1005.
13. Paetau, I., M. S. Thesis "Biodegradable plastics made from soy protein." Iowa
     State University, Ames, Iowa. 1993.

Journal Paper No. J-15513 of the Iowa Agriculture and Home Economics Experiment
Station, Ames, Iowa Project No. 2863.

RECEIVED May 24, 1994

Chapter 7

Formulations of Polyurethane Foams with Corn Carbohydrates

R. L. Cunningham[1], M. E. Carr[1], E. B. Bagley[1], S. H. Gordon[2], and R. V. Greene[2]

[1]Food Physical Chemistry Research Unit and
[2]Biopolymer Research Unit, National Center for Agricultural Utilization Research, Agricultural Research Service, U.S. Department of Agriculture, Peoria, IL 61604

The addition of pregelatinized corn flour or cornstarch to formulations of hydrophilic foams provides filled foams that exhibit properties of horticultural interest. The carbohydrate-filled polyurethane foams were formed by reacting a polyisocyanate-terminated polyether with water. The addition of 20% pregelatinized corn flour (prepolymer basis) to the formulation increased the volume of cured foams by 16%, whereas the same addition level of pregelatinized cornstarch increased the volume by 6%. Cornstarch or corn flour in the formulation produced foams requiring at least twice the force for 50% deflection compared to control foams. Water suction and drainage times were lower for the foams containing corn flour than those for either the controls or the foams containing cornstarch. However, water retention was lowest for the corn flour-filled foams.

Polyurethane (PU) markets are largest in transportation, furniture, and construction industries. These markets consumed 60% of the over 1.4 billion kg (3.2 billion pounds) of PU produced in the United States and Canada during 1991 (*1*). PU use has remained strong in 12 traditional markets and has penetrated new ones since 1989. Included in the market survey for the Polyurethane Division of The Society of the Plastics Industry, Inc. were flexible and rigid PU foams. In addition to the above markets, there is a growing interest in hydrophilic foams for a variety of specialty applications (*2*). These foams are used as absorbent materials in cleaning pads, disposable diapers, medical supplies, sanitary napkins, sponges, and horticultural products. Hydrophilic-type PU prepolymers are composed of polyisocyanate-terminated polyethers. Polyisocyanates contain more than one isocyanate (-N=C=O). These prepolymers are derived from toluene diisocyanate (TDI). The addition of water to the prepolymer forms an unstable carbamic acid (RNCO + H_2O → RNHCOOH), resulting in an amine and CO_2 (RNHCOOH → RNH_2 + CO_2).

The amine reacts with the polyisocyanate-terminated polyether to form urea chain extension and crosslinked products (RNH_2 + RNCO → RNHCONHR). Considerably more water can be used in hydrophilic foam formulations than in the formulations for the rigid PU foams with closed cells containing a blowing agent of a halogenated alkane to attain a low k-factor or thermal conductivity. Hydrophilic foams permit the blending of the corn carbohydrates in water before adding them to the other ingredients and moisture content does not limit the quantities of the carbohydrates that can be used.

Previously, we prepared and evaluated rigid PU foams containing unmodified, high amylose, or waxy cornstarches, and unmodified or pregelatinized corn flour (10%, by weight) (3). Of the five levels of unmodified corn flour added to the formulations of rigid foams, 10% corn flour (polyol basis) in the formulation maintained foam compressive strength and insulating properties (4). Incorporation of 20% canary dextrin (polyol basis) into rigid PU formulations provided foams with similar or improved thermal stability, open-cell content, and thermal conductivity as compared to the controls not containing dextrin (5). Substitution of 10% of the polyether polyol with an equal weight of cornstarch gave foams exhibiting 4% less compressive strength but similar insulating values compared to the control foams (6).

Recently, unmodified cornstarch and corn flour were incorporated into hydrophilic PU foam formulations (7). In this study, hydrophilic foams containing pregelationized corn carbohydrates (flour and starch) were prepared and evaluated for their physical properties and solubilities of the corn carbohydrate components in an amylolytic bacteria culture.

Experimental

Materials. Hypol FHP 3000 polymer, toluene diisocyanate-terminated polyether, was supplied by Hampshire Chemical Corp., Lexington, MA. This prepolymer contains less than 9% excess TDI. A laboratory hood is required in working with the polymer to provide adequate ventilation. TDI can cause severe allergic respiratory reaction; eye, skin, and respiratory irritation; allergic skin reaction; and is listed on the National Toxicology Program (NTP) carcinogen list as a suspect carcinogen based on animal studies. BRIJ 72, surfactant, was supplied by ICI Americas Inc., Wilmington, DE. The pregelatinized corn products, flour and starch, added to the formulations were Product 961 and Amaizo 2414 supplied by Illinois Cereal Mills, Inc., Paris, IL and American Maize-Products Co., Hammond, IN, respectively.

Foam Preparation. Foams were formulated using 100 g of Hypol FHP 3000 polymer with the ingredients listed in Table I. The 20% addition level of pregelatinized products was selected on the basis of the viscosity of the product in water so that ease of mixing could be achieved. Higher levels of gelatinized product in the formulation would require additional water. Four replicate foams were prepared with each formulation. BRIJ 72, surfactant, was added to water and mixed with a model TS 2010, Lightning mixer (General Signal Corp., Rochester, NY) with A-310 impeller (6.35 mm) for eight min at 500 rpm in a

1.1-L container. The cornstarch or corn flour component was added to the surfactant/water mixture and blended using a glass stirring rod. This mixture was added to the Hypol polymer and blended with a glass rod. The ingredients in the control foams required 1 1/2 min of mixing to achieve creaming (visual change) as compared to one min with addition of the corn products. The ingredients were poured into wooden boxes (178 X 178 X 76 mm) and allowed to rise at room conditions. After 12 h, foams were placed in a Precision Scientific STM 135 Mechanical Convention Oven (Precision Scientific, Chicago, IL) at 70°C for 14.5 h.

Physical Testing. Foam specimens with heights of 25 mm were prepared using a band saw. The specimens were conditioned for 12 h at 23°C and 50% relative humidity. Densities were determined on unpreflexed foams by procedures of The American Society for Testing and Materials (ASTM) procedure D 3574-86, Test A. Each foam was cut into four specimens of 51 X 51 X 25 mm with a band saw. Compression force deflection tests were performed on two of these specimens from each of four replicate foams according to ASTM D procedure 3574-86, Test C using an Instron apparatus (model 4201, Instron Corp., Canton, MA) equipped with a 5-kN static load cell, type 2518-805. A suspended, self-aligning pressure pad was mounted under the cross arm for the loading platen. These specimens were used also for the tests of solubilization of carbohydrate in media as described below.

Solubilization of Carbohydrate in Media. Foams containing no additives (control) and those containing either pregelatinized cornstarch or pregelatinized corn flour were shredded into 3-mm pieces to increase surface availability. Each sample was weighed and placed into a 250-mL Erlenmeyer flask and autoclaved at 121°C (250°F) for 15 min at 0.103 MPa (15 psi). After being autoclaved and cooled to room temperature, 100 mL of a modified basic media (8) was added. The *Spirillum* Nitrogen-Fixing Medium was modified for bacteria by addition of ammonium chloride and vitamin solution, whereas sodium malate was omitted. Control flasks were placed on an orbital shaker (Lab-Line Instruments, Inc., Melrose Park, IL) at 28°C and 180 rpm. To the test flasks, two mL of LD 76 (9) (a proprietary mixed bacterial culture composed of amylolytic bacteria) was added to each flask and the flasks placed on the orbital shaker. After 28 days, samples were removed from the shaker and centrifuged at 15,300 X g for 15 min with model J2-21, JA-14 rotor (Beckman Instruments, Inc., Palo Alto, CA). Supernatant was drawn off and the solids were resuspended in sterile wash media. The wash procedure was repeated three times. The third wash of the samples was with sterile water. Supernatants were filtered through disposable filterware (Type S cellulose nitrate, 0.2 micron pore size, Nagle Co., Subsidiary of Sybron, Rochester, NY). Each sample including solids on filters was placed in a dish, dried in an oven at 80°C overnight, and weighed.

Hydrophilicity. Tests were performed on two specimens from each of the four replicate foams. Procedures for determining the suction and drainage times of foams were similar to those described in *Polyurethane Handbook* (10). The 51

Table I. Formulation of Ingredients

Ingredients	Grams
Hypol prepolymer	100
Water	200
BRIJ 72, surfactant	1.5
Pregelationized corn products	20
(Cornstarch or corn flour)	

Table II. Foams with Pregelationized Products

Properties	Control	Cornstarch[a]	Corn flour[a]
Density, (g/cm^3)	0.157	0.178	0.162
Compression force			
Thickness, mm[b]	28.6	27.9	26.9
Deflection 50%, Pa	15,000	35,700	30,100
Hydrophilicity			
Time, s			
suction	729	988	407
drainage	102	106	74.9
Increase, %			
weight	152	102	80.3
density	89.1	63.8	46.2
volume	33.0	23.6	23.3

[a] Sixteen percent, dry weight basis.
[b] After contact load.

X 51 X 25-mm specimens were immersed in a pan containing a 24-mm depth of water. Time required to saturate the foam with water until the upper surface was wetted is called the suction time. Saturated foam was removed from the pan and placed in a beaker at an angle for the foam to drain. The water draining time was from the time the specimen was removed from the water to the completion of draining (no dripping for one min).

Scanning Electron Microscopy (SEM). Shredded specimens were mounted on aluminum stubs using double-sided carbon tape and were coated with gold-palladium (60:40) to a thickness of about 0.015 micron in a sputter coater. The coated specimens were observed in a SEM model JSM-6400 (JEOL Inc., Peabody, MA) at a specimen angle of O°. Accelerating voltage was 10 KV, and final aperature was 200 microns.

Fourier Transform Infrared (FTIR) Spectrometric Analysis. Foam samples were dried under vacuum at 60°C for 24 h. Samples were ground, mixed with KBr, and pressed into transparent KBr disks. Particle size of the powders was minimized to give homogenous KBr disks. This was accomplished by pulverizing 5.0 mg of sample for three min at liquid nitrogen temperature in a stainless steel vial containing two stainless steel ball-bearings on a mixer mill (Brinkmann Instruments Inc., Subsidiary of Sybron Corp., Westbury, NY). After warming to ambient temperature, 95.0 mg of spectral grade KBr (Spectra-Tech Inc., Stamford, CT) was added to the vial. All weighings and transfers of samples were done in a dry box to prevent moisture absorption by the hygroscopic KBr. The sample in KBr was pulverized on an amalgamator (Wig-L-Bug, Crescent Dental Mfg. Co., Lyons, IL) for 60 s at liquid nitrogen temperature in the same vial. At ambient temperature, 25 mg of the pulverized KBr mixture was diluted to 750 mg in KBr and pulverized on the amalgamator. Three hundred mg of the pulverized KBr mixture was transferred in the dry box to a 13-mm KBr die (Perkin-Elmer Corp., Analytical Instruments, Norwalk, CT), and the box was evacuated for five min before pressing *in vacuo* at 110 MPa on a laboratory press (Fred S. Carver, Menomonee Falls, WI). Infrared spectra were measured on an FTIR spectrometer (model RFX-75, KVB-Analect, Irvine, CA) equipped with a TGS detector. Interferograms were processed on an APT-824 array processor using triangular apodization for linear response. Spectra were acquired at 4 cm^{-1} resolution and signal averaged over 32 scans with no zero filling. The interferometer and sample chambers were purged with dry nitrogen to remove spectral interference from water vapor and carbon dioxide.

Results and Discussion

Physical Tests. Foams containing cornstarch (20%, prepolymer basis, Table I) exhibited higher densities than those containing the same amount of corn flour or no additives. Control samples were difficult to cut with the band saw because of their softness. Compression force deflection tests measure the force necessary to produce a 50% compression over the entire top area of the foam specimens. As shown in Table II, the addition of 20% cornstarch or corn flour to the formulations produced foams requiring at least twice the force for 50% deflection providing firmer foams. One-fifth additional force was required with the foams containing cornstarch. Polymer composition, density, and cell structure and size contribute to compressive behavior (11). Dry volumes of foams containing additives were larger than those of foams containing no additives (Figure 1). Final volumes of foams containing 16% (dry basis) corn flour were 16% larger and those foams containing cornstarch were 6% larger than those of the control foams. Additives reduce shrinkage in linear dimensions during drying. The use of these pregelatinized corn products permits foam volume expansion with acceptable physical properties.

Hydrophilicity. The suction and drainage times were lower for the foams containing corn flour than those for the controls or foams containing cornstarch (Table II). Controls and cornstarch foams had similar drainage times. After saturation and drainage, the weights of the foams containing 16% pregelatinized cornstarch were greater than for the foams containing the same amount of pregelatinized corn flour. After drainage, volume increases were the same for the foams containing additives but less than those of the controls.

Solubility of Carbohydrate in Media. In the United States and other countries, local and national legislative activity has resulted in support of research and development on biodegradable replacements for current plastics (12). The dry and wet milling of corn provides a flour or starch that is abundantly available at a relatively low price. These products can serve as fillers and biodegradable additives to plastics. With the prepolymer containing only 9% excess TDI, most of the TDI would react with the water and not with the corn carbohydrate. With the addition of pregelatinized corn flour or cornstarch to cellular plastics, the data in Table III show weight losses after 28 days of incubation equal to the amount of carbohydrate additive. The polyurethane segment is not biodegradable. Weight losses of shredded foams containing corn products without added bacteria indicate some solubility of the corn products in the media and suggest the presence of some endogenous microbes in the foams that are difficult to kill using standard sterilization procedures.

The SEM photomicrographs of the foams before and after treatment with the consortium of bacteria (LD 76) are shown in Figure 2 (A & B). Good incorporation of the additives into the foams is observed. FTIR spectra of the corn flour-filled foam before and after the bacterial treatment are shown in Figure 3 (A & B). The presence of flour is noted by the C-O absorption band in the carbohydrate fingerprint region (1000-1200 cm^{-1}) before the bacterial treatment (Figure 3A, see arrow) and its absence after 28 days of incubation (Figure 3B, see arrow). These spectra correspond to the weight data (Table III) that show the loss equal to the cabohydrate additive in the foam.

Conclusions

The addition of 20% (based on weight of Hypol prepolymer) of pregelatinized cornstarch and corn flour to a hydrophilic foam formulation produced foams with

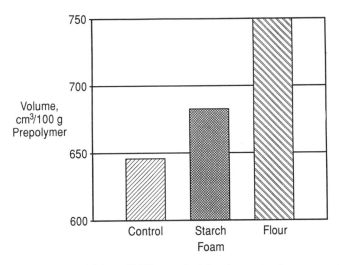

Figure 1. Effect of additives (20% pregelatinized cornstarch or corn flour, prepolymer basis) on foam volumes.

Table III.   Effect of Mixed Bacterial Culture
on the Weight Loss (%) of Foam[a]

Days	(Foam, 3mm)	Additives		
	None	Cornstarch[b]	Corn flour[b]	
28	1.0	15	20
28[c]	3.8	16	18

[a]   Incubated at 28°C in a liquid culture (LD 76).
[b]   Pregelationized product, 16% dry weight basis.
[c]   Not inoculated.

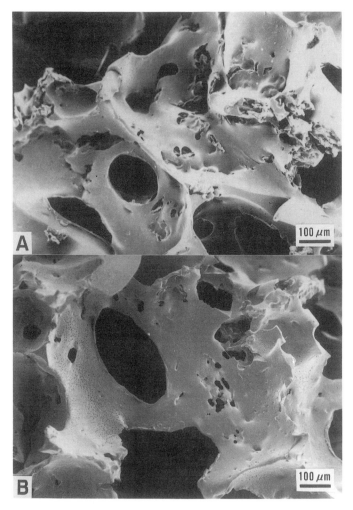

Figure 2. Scanning electron microscopy photomicrographs of hydrophilic foam (3-mm pieces) before (A) and after (B) 28 days of incubation with a mixed bacterial culture. Additive: 20% corn flour.

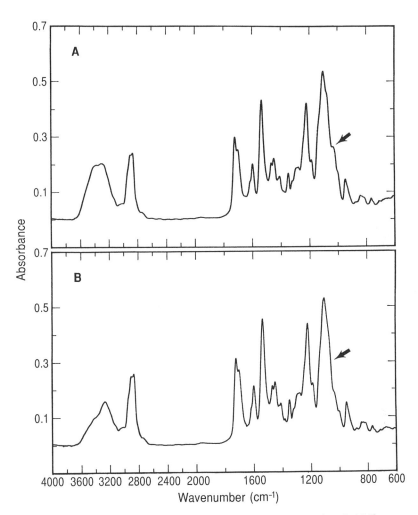

Figure 3. Fourier transform infrared spectrometric analysis of 16% corn flour-filled foam (3-mm pieces) before (A) and after (B) 28 days of incubation with a mixed bacterial culture.

increased volumes and greater resistance (at least 2X) to 50% deflection. Suction and drainage times were faster with the corn flour-filled foams than with the cornstarch-filled foams.   After 28 days of incubation with a consortium of amylolytic bacteria, weight loss determinations and FTIR spectra indicate the absence of the corn carbohydrate additives. These additives extend foams and provide a new type of product with good hydrophilicity characteristics for potential horticultural uses.

Acknowledgments

The authors thank F. L. Baker, G. D. Grose, C. James, and M.
P. Kinney for their technical contributions.
* The mention of firm names or trade products does not imply that
they are endorsed or recommended by the U. S. Department of
Agriculture over other firms or similar products not mentioned.

Literature Cited

1. *Polyurethanes 92 Post Conference Report*; Polyurethane Division,
    The Society of the Plastics Industry, Inc.: New York, NY, **1992**;
    p 1.
2. Frisch, K. C. In *International Progress in Urethanes*; Frisch,
    K. C. and Hermandey A. Eds., Technomic Pub,: Lancaster, PA,
    1977; pp 12-33.
3. Cunningham, R. L.; Carr, M. E. *Corn Utilization Conference
    III Proc.*; National Corn Growers Association and Ciby-Geigy
    Seed Division: St Louis, MO, **1990**; pp 424-439.
4. Cunningham, R. L.; Carr, M. E.; Bagley, E. B. *Cereal Chem.*
    1991, *68*, 258-261.
5. Cunningham, R. L.; Carr, M. E.; Bagley, E. B. *J. Appl.
    Polym. Sci.*, **1992**, *44*, 1477-1483.
6. Cunningham, R. L.; Carr, M. E.; Bagley, E. B.; Nelsen, T. C.
    Starch/Stärke **1992**, *44*(4), 141-145.
7. Cunningham, R. L.; Carr, M. E.; Bagley, E. B.; Gordon, S. H.;
    Greene, R. V. *J. Appl. Polym. Sci.*, **1994**, *51*, 1311-1317.
8. Atlas, R. M. In *Handbook of Microbiological Media*; Parks, L.
    C., Ed.; CRC Press, Inc.: Boca Raton, FL, **1993**; p 826.
9. Gould, J. M.; Gordon, S. H.; Dexter, L. B.; Swanson, C. L. In
    *Agricultural and Synthetic Polymers; Biodegradability and
    Utilization*; Glass, J. E.; Swift, G. Eds.; ACS Symp. Ser. No.
    433; American Chemical Society: Washington, DC, **1990**; pp 65-75.
10. Lowe, G. In *Polyurethane Handbook*; Oertel, G., Ed.; Carl
    Hanser Verlag: Munich, Germany, **1985**, pp 309-310.
11. Suh, K. W. In *Polymeric Foams*; Klempner, D.; Frisch, K. C.,
    Eds.; Carl Hanser Verlag: Munich, Germany, **1991**, p 162.
12. Doane, W. M. *New Crops, New Uses, New Markets*; U. S. Department
    of Agriculture, U. S. Government Printing Office: Washington,
    DC, **1992**, pp 147-153.

RECEIVED May 24, 1994

POLYSACCHARIDES, OLIGOSACCHARIDES, AND MONOSACCHARIDES

Chapter 8

Hydrogel-Type Polymers in the Controlled Release of Phytodrugs

R. Solaro[1], S. D'Antone[1], E. Chiellini[1], A. Rehab[2], A. Akelah[2], and R. Issa[2]

[1]Department of Chemistry and Industrial Chemistry, University of Pisa, Via Risorgimento 35, 56100 Pisa, Italy
[2]Department of Chemistry, University of Tanta, Tanta, Egypt

The preparation and structural characterization of polymeric herbicides consisting of 2,4-dichlorophenoxyacetic acid and 4-chloro-2-methyl–phenoxyacetic acid either covalently or ionically bound to linear and crosslinked polymers are described. Poly(styrene/divinylbenzene) resins, crosslinked polysaccharides, and homo and copolymers of oligo(oxy-ethylene) monomethacrylates were used as polymeric supports Herbicide binding was attained by nucleophilic displacement, esterification and ion-exchange reactions. Herbicide release from polymer beads was monitored in water solution buffered at pH 4, 7 and 9. The observed release profiles are discussed in terms of polymer inherent structural features. The release kinetics did not fit simple diffusional schemes while could be satisfactorily reproduced by the contemporary occurrence of two exponential decay processes, differing by two orders of magnitude in their rates.

Utilization of chemicals in agriculture, as administered by conventional routes, is often wasteful and moreover can cause serious environmental problems, harmful to wildlife and even to humans. Delivery of agrochemicals by controlled release formulation is more than a promising concept offering advantages of ecological and economical relevance (1–4). The replacement of conventional herbicide application by controlled release techniques would avoid spreading an excess of toxic substances and at the same time could increase the efficiency of administration. Synthetic and naturally occurring polymers are the cornerstone of this technique.

Hydrophilic polymers and polymeric hydrogels have been largely studied and applied in several fields and they appear to deserve attention for the formulation of controlled release systems to be specifically used in sandy soils (5–10). Indeed systems displaying the capability of retaining a fairly large amount of water might help to diminish the pressure for frequent irrigation and hence washing out of any applied phytodrug to that kind of soils.

0097–6156/94/0575–0112$08.00/0

In this connection and within the framework of an international cooperation purposely established, we have undertaken a research project focused on the preparation of functional polymeric hydrogels of synthetic and semisynthetic origin, designed to bind either covalently or ionically powerful biocides (*11–16*). They should possess specific properties able to guarantee for a sustained controlled–release of active agents, tuneable according to external parameters (pH and/or saline concentration) and at the same time capable of retaining a fairly large amount of water even under severe temperature and soil conditions. Moreover, controlled release formulations based on suitably modified natural or semisynthetic polymers are of special interest, because of their relative low cost and biodegradability properties.

Synthesis of Polymer Supported Herbicides

Hydrogels containing oligo(oxyethylene) moieties were prepared by radically initiated polymerization of oligo(oxyethylene) methacrylates (degree of oligomerization n = 2, 4 and 8), alone or in the presence of N,N–methylenebisacrylamide (MBAA) as a crosslinking agent.

The resulting hydrogel–type polymers were reacted in the presence of triethylamine with the chlorides of 2,4–dichloro-phenoxyacetic acid (2,4–D) and 2–methyl–4–chlorophe-noxyacetic acid (MCPA) to afford covalently bound polymeric herbicides (**P1**). The herbicide loading, evaluated by saponification of the ester group and subsequent titration, was between 5 and 55%–mol whereas the swelling in water ranged between 50 and 400% (Table I).

An analogous series of polymeric herbicides having a comparatively higher herbicide content were prepared by direct homopolymerization and copolymerization of herbicide monomers obtained by reacting oligo(oxyethylene) monomethacrylates with the acid chloride of MCPA and 2,4–D.

2,4-D   R = Cl
MCPA   R = CH$_3$

The swelling in water of the polymer samples obtained by direct polymerization of herbicide containing methacrylates ranged between 10 and 140% (Table I). As expected the copolymer samples containing hydrophilic co–units exhibited the largest values. The increase of the length of the oligo(oxyethylene) groups produced a slight increase of the water uptake, whereas an increase of the degree of crosslinking reduced the swellability of the samples. In any case these values were much lower than those observed for the corresponding samples obtained by post–reaction on preformed poly[oligo(oxyethylene) methacrylate]s, having a lower herbicide loading. This behavior can be attributed to the strong hydrophobicity of herbicide residues.

P1   R = Cl, CH$_3$

Some hydrogel polymers with modulated hydrophilicity (swelling in water included between 150 and 750%) were prepared by copolymerization of tetraethylene

Table I. Polymeric Herbicides (P1) Obtained by Covalent Binding of Herbicides on Oligo(oxyethylene) Methacrylates

n	MBAA (mol–%)	Herbicide [a] type	(mol–%)	Swelling [b] (%)	Release [c] pH 4	pH 7	pH 9
2	0	2,4–D	22	90	12	16	45
2	0	"	100	10	1	2	43
2	5	"	25	90	7	7	32
2	10	"	27	50	12	13	37
4	0	"	15	110	8	14	88
4	0	"	100	31	1	2	11
4	5	"	11	70	12	20	54
4	10	"	24	70	13	25	59
8	0	"	45	410	19	18	81
8	0	"	100	42	10	10	74
8	5	"	17	370	38	43	91
8	10	"	22	290	26	25	75
2	0	MCPA	34	120	1	4	n.d.
2	0	"	100	13	1	1	2
4	0	"	100	36	2	3	5
4	5	"	93	32	2	2	12
4	10	"	87	20	2	2	15
8	0	"	100	48	16	20	68
2	5	"	5	100	24	34	86
4	0	"	18	130	4	27	53
4	5	"	25	120	4	31	52
4	10	"	19	120	4	21	47
8	0	"	47	210	31	36	88
8	5	"	34	140	25	33	89
8	10	"	50	130	20	23	54

[a] Evaluated by elemental analysis and by saponification. [b] Evaluated as weight of sorbed water per 100 g of dry polymer. [c] Percent herbicide released after 20 days.

glycol methacrylate with acrylamide (AAm). The resulting copolymers were reacted with 2,4–D acid chloride in the presence of triethylamine to give the corresponding polymeric herbicides (**P2**, R' = $CONH_2$).

An analogous series of polymeric herbicides with different hydrophilicity (**P2**) was also prepared by copolymerization of 4–chloro–2–methylphenoxyacetyl-tetra(oxyethylene) methacrylate with different comonomers, such as acrylamide, 4–vinylpyridine (VP), di(oxyethylene) monomethacrylate (DEGMA), and octa(oxyethylene) monomethacrylate (OEGMA). All the samples prepared resulted insoluble in water and in common organic solvents, this was very likely due to the presence of a small amount of dimethacrylate moieties in the starting monomers. The swelling in water of these polymeric matrices, evaluated as weight of absorbed water per 100 g of dry polymer, ranged between 45 and 170. As expected the copolymer samples containing more hydrophilic co–units exhibited the largest values, whereas an

Table II. Polymeric herbicides (P2) obtained by covalent binding of herbicides on copolymers of tetra(oxyethylene) methacrylate with different comonomers

Comonomer		Herbicide [a]		Swelling [b]		Release [c]	
type	(mol–%)	type	(mol–%)	(%)	pH 4	pH 7	pH 9
AAm	25	2,4–D	31	130	6	20	57
AAm	50	2,4–D	27	170	13	n.d.	40
AAm	75	2,4–D	13	130	28	n.d.	76
AAm	25	MCPA	74	46	16	16	75
AAm	50	MCPA	46	85	7	9	77
AAm	75	MCPA	24	142	3	6	94
VP	50	MCPA	45	53	25	20	80
DEGMA	50	MCPA	46	64	1	3	77
OEGMA	50	MCPA	50	126	13	16	84

[a] Evaluated by saponification. [b] Evaluated as weight of sorbed water per 100 g of dry polymer. [c] Percent herbicide released after 20 days.

increase of the content of hydrophobic herbicide moieties reduces the swellability (Table II).

Up to now, very few data have been reported on the formulation of polymeric pro–herbicides in which the active ingredient is ionically held into the polymer matrix (*10*). Ionic binding of MCPA on preformed homopolymers and copolymers of tetra(oxyethylene) methacrylate with MBAA and AAm was carried out according to the following reaction sequence. The terminal hydroxyl groups in oligo(oxyethylene) side chains were first chloroacetylated by reaction with chloroacetyl chloride in the presence of triethylamine. The observed degree of esterification ranged between 12 and 95%, higher values being reached at lower crosslinking densities and longer side chains. The polymeric chloroacetyl derivatives were then reacted with excess tri–*n*–butylamine [Y = N(C$_4$H$_9$)$_3$] or triphenylphosphine [Y = P(C$_6$H$_5$)$_3$] to form the corresponding quaternary onium salts. The observed conversions, in the 30–90% range, were independent of both polymer structure and type of nucleophile used. The functionalized polymer matrices were finally loaded with 4–chloro–2–methyl-phenoxyacetic acid by ion exchange reaction with the sodium salt of MCPA in water/methanol (1:9 v/v) suspension. By following this route, polymeric herbicides having an overall MCPA content in the range 0.1–0.5 mmol/g were obtained (Table III).

All polymer samples containing ionically bound herbicide (**P3**) displayed a moderate to strong hydrogel character (degree of swelling 130–600%). At comparable quaternization extent, the swelling in water of polymeric herbicides was dependent upon both length of the oligo(oxyethylene) segment and degree of crosslinking. The

Table III. Polymeric herbicides (P3) obtained by MCPA ionic binding on preformed homo and copolymers of tetra(oxyethylene) methacrylates

n	Comonomer type	(mol–%)	Q units[a] type	(mol–%)	MCPA[b] (mol–%)	Swelling[c] (%)	Release[d] pH 4	pH 7	pH 9
2	–	0	N	35	14	160	2	2	27
2	MBAA	5	N	10	9	133	6	5	35
2	MBAA	5	P	24	12	140	9	9	24
4	–	0	N	39	6	340	33	42	41
4	MBAA	5	N	11	6	270	8	9	28
4	MBAA	10	N	11	6	240	47	54	54
4	–	0	P	42	7	300	30	39	45
4	MBAA	5	P	35	5	200	12	16	31
4	MBAA	10	P	42	7	210	16	29	23
8	–	0	N	31	16	410	4	3	8
8	MBAA	5	N	16	16	300	50	46	48
8	MBAA	5	P	37	11	260	22	31	40
4	AAm	25	N	17	13	430	6	7	25
4	AAm	50	N	13	3	540	7	8	25
4	AAm	75	N	8	3	580	42	57	53
4	AAm	95	P	40	6	480	14	10	15

[a] Quaternized units. [b] By elemental analysis and by saponification. [c] Evaluated as weight of sorbed water per 100 g of dry polymer. [d] Percent herbicide released after 100 days.

highest values were attained by AAm copolymers, whose water uptake was driven more by their AAm content than by type of quaternary onium group or extent of quaternization.

Two samples of styrene/divinylbenzene resins containing 2 and 12.5 %–mol of DVB, respectively were chloromethylated (degree of chloromethylation > 90%) by reaction with chloromethyl methyl ether in the presence of $ZnCl_2$. In order to increase the resin hydrophilicity, chloromethyl groups were partially reacted with either oligo(ethylene glycol)s, having degree of oligomerization 6, 9 and 13, or the corresponding monomethyl ethers to yield styrene/DVB resins containing 14–55 %–mol of oligooxyethylenated units and 72–22 %–mol of chloromethylated units.

Chloromethylated styrene/divinylbenzene resins were covalently loaded with MCPA by nucleophilic displacement of the chloride of the chloromethyl groups with MCPA sodium salt to give polymeric herbicides **P4**. The resins loading was attained by esterification of the pendant hydroxyl groups with the acid chlorides of 2,4–D and

$\sim\sim CH_2-CH\sim\sim CH_2-CH\sim\sim CH_2-CH\sim\sim$

P4

MCPA in the presence of triethylamine (**P5**). In all cases multifunctional polymer–bound herbicides having an herbicide content ranging from 0.4 to 1.4 mmol per gram of resin were obtained in fairly high yields (Table IV).

In order to obtain ionically bound polymeric herbicides, chloromethylated styrene/DVB resins containing monomethoxyoligo(oxyethylene) glycol residues were reacted with excess tri–n–butylamine to yield styrene resins containing oligo(oxyethylene) and quaternary ammonium groups. Loading with MCPA biocide was then performed by an ion exchange procedure. By this route ionically bound polymeric herbicides (**P6**) containing 0.4 to 0.7 mmol of herbicide per gram of dry resin were obtained (Table IV).

Finally a series of polymeric herbicides based on crosslinked poly(saccharide)s (**P7**) was prepared by reacting alkaline water solutions of semisynthetic polymers, such as hydroxyethylcellulose (HEC) and dextranes (D4 and D7, Mn $4 \cdot 10^4$ and $7 \cdot 10^4$,

Table IV. Polymeric herbicides (P4–P6) obtained by covalent or ionic binding of herbicides on styrene/DVB resins

DVB (mol-%)	n	R	OE units [a] (mol-%)	QA units [b] (mol-%)	Herbicide [c] Type	(mol-%)	Release [d] pH 4	pH 7	pH 9
2.0	6	H	40	0	2,4–D	36	3	7	11
12.5	6	H	56	0	"	32	5	15	24
2.0	6	H	40	0	MCPA	37	6	15	29
2.0	9	H	32	0	"	29	14	12	90
12.5	6	H	56	0	"	14	13	30	48
12.5	9	H	14	0	"	8	15	32	61
2.0	6	H	40	0	"	36	6	14	25
2.0	6	CH$_3$	20	0	"	28	7	12	37
2.0	9	CH$_3$	18	0	"	18	17	30	79
2.0	13	CH$_3$	55	0	"	55	11	28	48
12.5	6	CH$_3$	34	0	"	42	4	14	59
2.0	6	H	40	22	"	21	80	80	100
2.0	6	CH$_3$	20	24	"	17	68	85	99
2.0	9	CH$_3$	18	10	"	9	89	82	100
2.0	13	CH$_3$	55	31	"	31	66	70	91
12.5	6	CH$_3$	34	37	"	16	46	58	66

[a] Oligo(oxyethylene) units. [b] Quaternized units. [c] Evaluated by elemental analysis and by saponification. [d] Percent herbicide released after 94 days.

$\sim\sim CH_2 - CH \sim\sim CH_2 - CH \sim CH_2 - CH \sim\sim$ $\sim\sim CH_2 - CH \sim\sim CH_2 - CH \sim CH_2 - CH \sim\sim$

P5   Cl   **P6**

respectively) with epichlorohydrine in toluene, under stirring and in the presence of suitable suspending agents. Molar ratios epichlorohydrine/glucose units (EP/GLU) ranging from 0.4 to 3 were used (Table V). The crosslinked polymers, isolated as small beads, were characterized by a degree of swelling included between 400 and 1200, thus substantiating a rather high hydrogel–forming capability for the prepared samples.

As expected, the degree of swelling of crosslinked samples based on HEC and D7 decreased with increasing the amount of epichlorohydrine in the reaction mixture, whereas almost constant values were recorded for samples derived from D4. This

Table V. Polymeric herbicides (P7) obtained by covalent binding of 2,4–D on polysaccharides crosslinked with epichlorohydrine

| Type | Starting polymer | | Polymeric pro–herbicide | | | | |
	EP/GLU [a] (mol/mol)	Swelling [b] (%)	Loading [c] (mol/mol)	Swelling [b] (%)	Release [d] pH 4	pH 7	pH 9
HEC	0	–	2.21	–	1	2	2
	0.44	1060	0.47	160	5	5	100
	3.00	750	1.03	170	24	21	63
D4	0	–	0.79	–	6	8	99
	0.72	615	0.54	70	21	14	98
	1.15	655	1.19	80	< 1	< 1	22
	1.44	515	1.53	10	< 1	< 1	6
D7	0	–	0.98	–	1	1	34
	0.72	1220	0.50	10	3	5	95
	1.15	815	0.96	10	4	6	69
	1.44	390	1.83	50	< 1	< 1	1

[a] Molar ratio epichlorohydrine/glucose units. [b] Evaluated as weight of sorbed water per 100 g of dry polymer. [c] Moles of bound 2,4-D herbicide per mole of glucose residue. [d] Percent herbicide released after 120 days.

seems to imply that the latter samples have almost the same crosslinking density, unless the lower molecular weight of the starting polymer attenuates the effect of the different amounts of crosslinking agent, by a simple molecular weight extension.

Crosslinked poly(saccharide)s were loaded with 2,4–D by direct esterification in DMSO solution, in the presence of carbonyl-diimidazole as coupling agent. For comparison the same reaction was also carried out on the starting uncrosslinked poly(saccharide)s. The resulting polymeric herbicides **P7** were characterized by a content of covalently bound herbicide ranging between 0.5 and 2.2 mol of 2,4–D residue per mol of glucose repeating unit (Table V). It is

interesting to note that the observed loading increases with increasing the EP content, thus suggesting that an appreciable amount of epichlorohydrine is grafted to the glucose residues as hydroxylated oligomeric moieties.

Herbicide Release

The release of herbicides from the above polymeric herbicides was investigated in buffer solution (pH 4, 7 and 9) at 25°C for time periods included between 20 and 120 days, depending on the overall release rate. In the case of samples containing ionically bound herbicides, a large excess of NaCl (20 g/l) was added to the buffer solution in order to provide a constant ion–strength and anion concentration throughout all release time. In any case, salting out effects of NaCl on the solubility of the released herbicides can be ruled out.

The amount of herbicide released within the time was monitored by UV analysis (Tables I-V). Typical examples of the dependence of kinetic profiles upon pH, spacer length, degree of crosslinking, polymer structure and chemical composition are represented in figures 1–6, in that order.

The reported data allow to draw the following considerations.

Independent of polymer structure and herbicide type, at pH 9 a fairly high release extent, most often higher than 40% was observed, whereas definitely lower values were attained at pH 7 and 4 (Fig. 1). The marked influence at even slightly alkaline conditions on the herbicide release rate is in accordance with the nature of the functional bond involved into the hydrolytic cleavage (*17, 18*). A less pronounced influence by pH on the release extent was observed for ionically bound herbicides (Fig. 2), which is in agreement with the occurrence of an ion–exchange release mechanism. The slightly higher release values attained at pH 9 suggest that some herbicide cleavage may take place also by hydrolysis of the quaternized acetate groups.

The release rate generally increased on increasing swelling in water and hence polymer hydrophilicity. In accordance, enhancement of both polymer hydrophilicity and conformational mobility by extending the oligo(oxyethylene) segment (Fig. 3), by

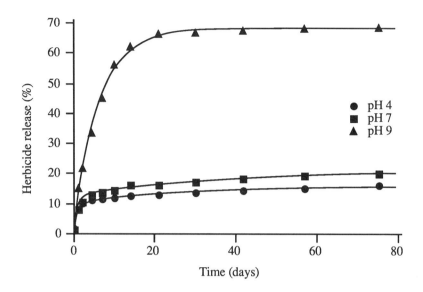

Figure 1. Influence of pH on the release of MCPA from **P1** polymeric herbicide (n=8, MBAA 0%, MCPA 48%)

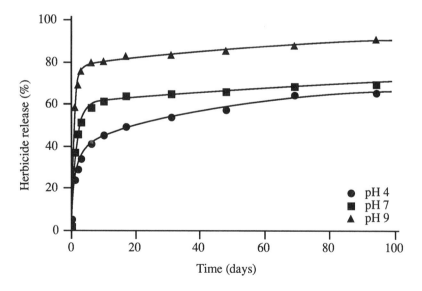

Figure 2. Influence of pH on the release of MCPA from **P6** polymeric herbicide (n=13, R=CH₃, MCPA 31%)

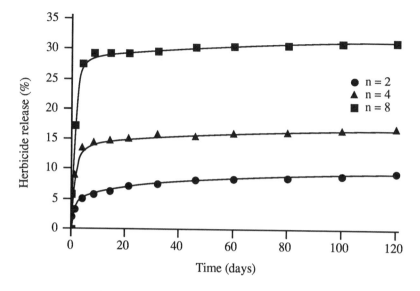

Figure 3. Influence of spacer length on the release of MCPA from **P3** polymeric herbicides (pH=7, MBAA 5%, type P)

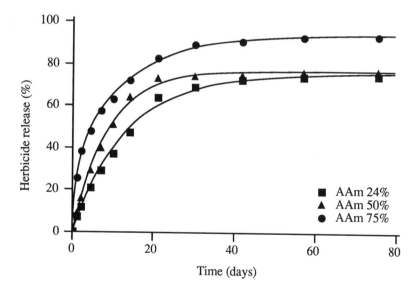

Figure 4. Influence of AAm content on the release of MCPA from **P2** polymeric herbicides (pH=9)

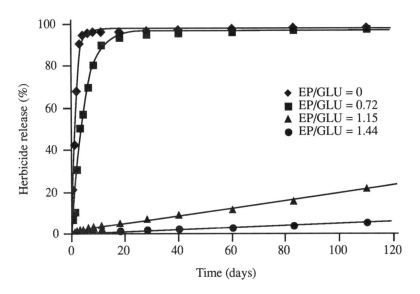

Figure 5. Influence of the degree of crossinking on the release of 2,4–D from **P7** polymeric herbicides (type D4, pH=9)

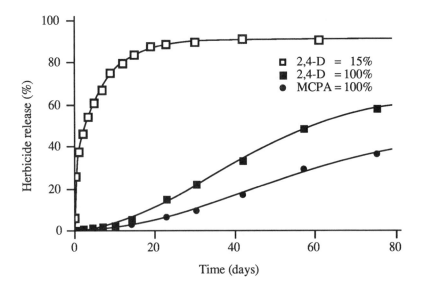

Figure 6. Influence of loading extent on herbicide release from **P1** polymeric herbicide (pH=9, n=4, MBAA 0%)

increasing the content of hydrophilic co–units (Fig. 4), and by lowering the degree of crosslinking (Fig. 5) gave rise to an increase of the release rate. Also the content of hydrophobic herbicide residues significantly affected herbicide release (Fig. 6). The induction period detected at higher loading can be attributed to the rather low hydrophilicity of the starting systems that partially prevented hydrolysis of the herbicide ester bond, whereas hydrophilicity and hence release rate increased as the hydrophobic herbicide groups were released (*19, 20*).

Independent of polymer structure, pH and herbicide type, the kinetic profiles were characterized by a much faster release rate at the early stages of release experiments and in any case release kinetics were not in agreement with simple diffusive or first order models. In most cases experimental data could be fitted by a biexponential kinetics, according to the following equation:

$$HR(t) = 100 - (X e^{-k_1 t} + (100-X) e^{-k_2 t})$$

where $HR(t)$ is the percent of herbicide released at time t, k_1 and k_2 are first order rate constants and X and $100-X$ represent the relative incidence of the two concurrent processes. Some typical experimental data (symbols) and relevant computed profiles (lines) are plotted in figures 1–7.

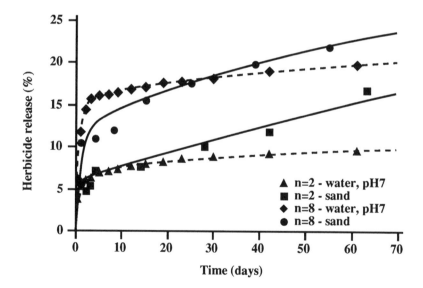

Figure 7. Release of 2,4–D from **P1** polymeric herbicides (MBAA 0%) in water and in sand.

Depending on polymer structure and pH, the computed k_1, k_2 and X values were often in the $1 \cdot 10^{-6}$–$2 \cdot 10^{-5}$ s^{-1}, $2 \cdot 10^{-10}$–$5 \cdot 10^{-8}$ s^{-1} and 1–55 range, respectively, the largest values being typically observed in pH 9 buffer solution. At present, no definite mechanism can be proposed to explain the origin and the different contributions of the two kinetic pathways operating at absolute rates differing by two–three orders of magnitude.

In order to establish the viability of the prepared polymeric herbicides for outdoor applications, the release of 2,4–D was preliminarily investigated in silica sand containing 50% water, taken as a simplified soil model. The observed release was comparable to that attained in water at pH 7 (Fig. 7). In spite of the crudeness of the model used, these data indicate that the investigated systems can be conveniently used in agricultural applications

Concluding Remarks

The present contribution, reporting on the work performed on the design of new polymeric matrices suitable to address the release of herbicides to draughty soils, can be considered as an effort aimed at providing further knowledge to a research area presenting yet a variety of pending issues.

Several routes for covalently or ionically supporting conventional herbicides on novel polymer matrices are described and the kinetics of their release in vitro, under different pH conditions, and in an experimental soil have been monitored. Both the overall extent of release and the kinetic profile are dependent upon the nature of the polymers, their swelling in water, and herbicide binding type.

In none of the investigated systems, however, the release kinetics fitted simple diffusional schemes. The kinetic profiles could indeed be reproduced by assuming the contemporary occurrence of two exponential decay processes, differing by at least two–three orders of magnitude in their absolute rates.

Acknowledgement

The financial support by MURST (60%) and by CNR – Italy (Progetto Finalizzato "Chimica Fine II") is gratefully acknowledged.

Literature Cited

1. *Controlled release pesticides*, Scher, H. B., Ed.; ACS Symp. Ser. No. 53; Am. Chem. Soc.: Washington D.C., 1977.
2. *Controlled release of pesticides and pharmaceuticals*; Lewis, D. H., Ed.; Plenum Press: New York, 1981.
3. *Polymers for biomedical and pharmacological applications*; Chiellini, E.; Giusti, P., Eds.; Plenum Press: New York, 1983.
4. *Reactive functionalyzed polymers: synthesis, properties and applications*; Akelah; A., Moet, A., Eds.; Chapmann & Hall: London, UK, 1989.
5. Allan, G. G.; Chopra, C. S.; Russel, R. M. *Int. Pest. Control.* **1972**, *14*, 15.
6. Allen, G. G.; Beer,J. W.; Cousin, H. J. *ACS Symp. Ser.* **1977**, *53*, 94.
7. McCormick, C. L.; Lichatowich, D. K.; Pelezo, J. A.; Anderson, K.W. *Polym. Prep. ACS Div. Polym. Chem.* **1980**, *21*, 109.
8. Connick, W. J., Jr. *Appl. Polym. Sci.* **1982**, *27*, 3341.

9. Pfister, G.; Bahadir, M.; Korte, F. *J. Controlled Release* **1986**, *3*, 229.
10. Shasha,B. S. In *High performance polymers: their origin and development*, Seymour, R. B.; Kirschenbaum, J. S., Eds.; Elsevier: Amsterdam, NL, 1986, p. 207.
11. Akelah, A.; Selim, A.; Hassaneim, M.; Rehab, A. *J. Polym. Mater.* **1985**, *2*, 149.
12. Rehab, A.; Akelah, A.; Issa, R.; Solaro, R.; Chiellini, E. *J. Controlled Release* **1990**, *13*, 1.
13. Solaro, R.; Chiellini, E.; Rehab, A.; Akelah, A.; Issa, R. *Reactive Polym.* **1991**, *14*, 21.
14. Rehab, A.; Akelah, A.; Issa, R.; D'Antone, S.; Solaro, R.; Chiellini, E. *J. Bioact. Compat. Polym.* **1991**, *6*, 52.
15. Issa, R.; Akelah, A.; Rehab, A.; Solaro, R.; Chiellini, E. *J. Controlled Release* **1991**, *17*, 113.
16. D'Antone, S.; Solaro, R.; Chiellini, E.; Issa, R.; Akelah, A.; Rehab, A. *New Polym. Mat.* **1992**, *3*, 223.
17. McCormick, C. L.; Kim, K. *J. Controlled Release* **1988**, *7*, 101.
18. McCormick, C. L.; Kim, K.; Ezzel, S. A. *J. Controlled Release* **1988**, *7*, 109.
19. Harris, F. W.; Aulabaugh, A. E.; Case, R. D.; Dykes, M. R.; Feld, W. A. *ACS Symp. Ser.* **1976**, *32*, 222.
20. Harris, F. W.; Dykes, M. R.; Baker, J. A.; Aulabaugh, A. E. *ACS Symp. Ser.* **1977**,

RECEIVED May 24, 1994

Chapter 9

Cornstarch-Derived Chemicals in Thermoplastic and Thermoset Applications

Larson B. Dunn, Jr.

Department of Food Science, University of Illinois
at Urbana–Champaign, Urbana, IL 61801

The cornstarch-derived chemicals described here are mainly D-glucose and methyl D-glucopyranosides. For thermoplastic applications, they were converted to polymerizable derivatives, such as allyl ethers and glucosides, acrylates and methacrylates, and were co-polymerized with methyl methacrylate or graft co-polymerized with poly(propylene), giving improvements in polymer properties, such as Tg, torsional modulus and modulus of elasticity. Mixtures of methyl glucosides were also used in thermoset applications, such as fiberglass insulation binders and plywood adhesives. In plywood adhesives, methyl glucosides showed remarkable formaldehyde scavenging properties in high formaldehyde resins. Similar properties were also demonstrated by methyl glucosides in other applications, such as textile finishing resins.

As the drought of the summer of 1988 demonstrated, even when yields are almost halved, there remains corn surpluses sufficient for all current needs. In normal years, American farms produce nearly twice what is needed in the United States or can be exported (1). Rather than penalize American farmers for having the highest productivity in the world, and since food uses for corn are growing only 3-4% annually, many organizations are funding research programs into non-traditional, non-food, or chemical, uses of corn-derived products (2). An example of this

0097–6156/94/0575–0126$08.54/0
© 1994 American Chemical Society

kind of program is the research being conducted in the Department of Food Science at the University of Illinois at Urbana-Champaign to incorporate cornstarch-derived chemicals such as glucose and methyl glucoside into a variety of thermoplastic and thermoset applications.

The benefits of research concerning the utilization of renewable, agricultural materials like those derived from cornstarch are many. With dwindling oil reserves, new sources of chemicals must eventually be found, and the farms of the midwest are an ideal source from an economic standpoint. From an American political viewpoint, with ever increasing petroleum imports from often unstable and volatile regions, it is desirable to have greater control over the source of important industrial materials, and again, the farms of the midwest are an extremely reliable source of products, in terms of both price and supply, even when there are severe droughts, as in 1988. With regard to health and safety, the use of renewable agricultural materials has definite benefits when compared to many potentially carcinogenic or flammable petroleum-based chemicals. Finally, there may be substantial environmental benefits in the use of degradable materials like cornstarch-derived chemicals, depending on the method of disposal, especially in light of recent concerns over filled-to-capacity landfills.

The basic philosophy that has guided this and other utilization research in the Department of Food Science at the University of Illinois is threefold. First and foremost, regardless of the type of agricultural material to be used or its application, an economic incentive must be provided to the end user. Industry generally desires to produce the same products it already produces at a lower cost. If there are additional benefits, such as health/safety or environmental, that is desirable, but not necessarily most important. Second, existing agricultural materials or products should be used with minimal modification to the materials or process in which they are to be used. In other words, the utilization process should be as simple as possible, both for economic reasons, as well as, to reduce the amount of change required by the agricultural producers and users. This will speed the success of the utilization process. Third, whatever the application being investigated, standard industry tests must be used to demonstrate the fitness of the agricultural material in the application. In other words, the language of the

industry must be spoken when talking to that industry about the
use of an agricultural material.

Processes to Produce Cornstarch-Derived Chemicals

The cornstarch-derived chemicals discussed in this chapter are
mainly the monosaccharides glucose and methyl glucoside (MeG).
D-Glucose, sometimes called dextrose, exists in solution in
mainly α- (1) and β-D-glucopyranose (2) forms, while the MeG
used in the utilization studies reported in this chapter

is generally a mixture of anomeric methyl α- (3) and β-D-
glucopyranosides (4), α- (5) and β-D-furanosides (6), higher
oligosaccharides (7) and some glucose (1, 2). They are produced,
respectively, by the hydrolysis and methanolysis of starch. The
hydrolysis of starch to produce glucose can be accomplished
either chemically, by simple acid hydrolysis, or enzymatically.
MeG is produced by the methanolysis of starch. Methyl α-D-
glucopyranoside (3) is often isolated as a pure, crystalline
product.

Thermoplastic Applications

Synthesis of Multi-functional Monomers (MFMs). The
rationale behind using cornstarch-derived materials like glucose
and methyl glucoside in thermoplastics is not to use them as
monomers to replace ethylene or propylene, and it is not to
produce a new, degradable plastic. The purpose in converting
glucose and methyl glucoside to polymerizable materials was to
produce multi-functional monomers (MFMs), polymer additives
usually based on polyols such as erythritol, glycerol or

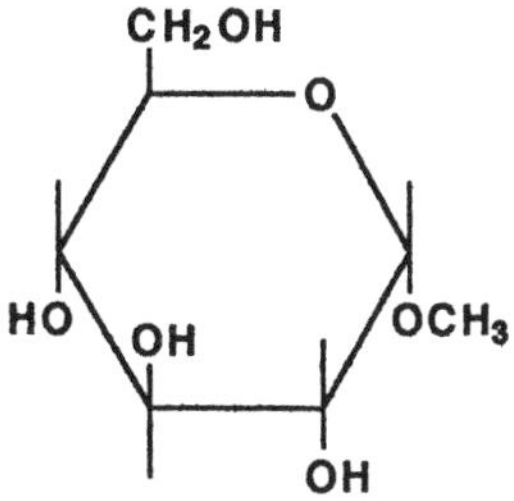

3

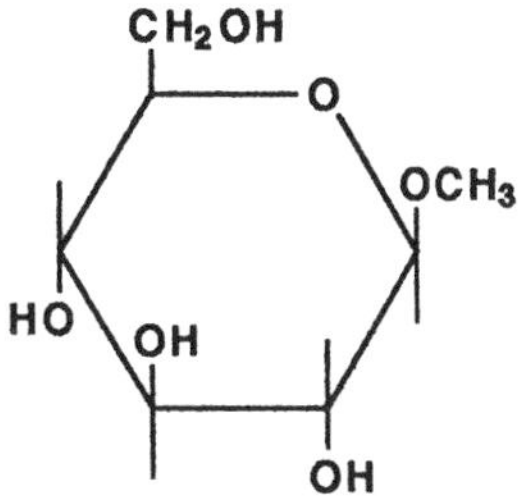

4

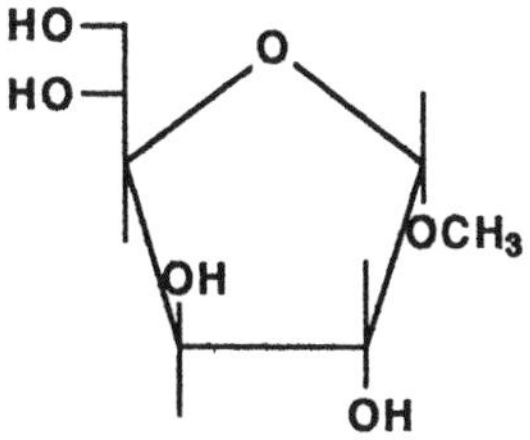

5

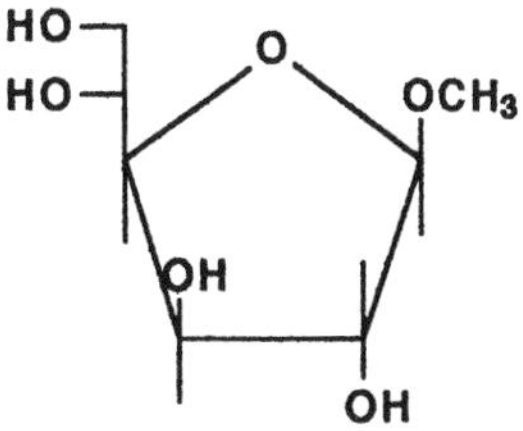

6

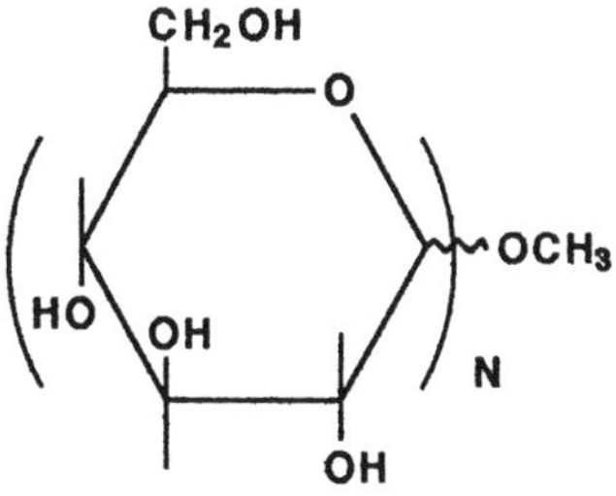

7

trimethylolpropane, that improve polymer properties through cross-linking. In addition, it was thought that MFMs from highly polar polyols such as glucose and MeG may make polymer formulations more compatible with extenders/fillers than commercial MFMs. Common extenders/fillers are usually salts, like calcium oxide, and the utility of such materials is generally limited by their lack of compatibility with non-polar polymers.

The cornstarch-derived MFMs were synthesized as quickly and simply as possible, as the object of this part of the research was to evaluate their performance as MFMs, not process development. The derivatives were analyzed by chromatography and by proton and carbon-13 nuclear magnetic resonance (NMR), and were generally used as isolated. In the course of the various syntheses, some interesting trends were observed.

The cornstarch-derived compounds and other carbohydrates for use in thermoplastic applications were chosen according to the classification in Table I. Three types of polymerizable derivatives were synthesized: esters, ethers and glycosides. In the process of synthesizing polymerizable glycosides from reducing sugars, those sugars were converted from reducing to non-reducing sugars.

Table I. Synthesis of Polymerizable Carbohydrate Derivatives

	Reducing	Non-Reducing
Monosaccharide	Glucose	Methyl Glucoside
Disaccharide	Lactose, Maltose	Sucrose

Ester and ether derivatives of MeG were first to be examined, as they are more easily synthesized. For thermoplastic applications, polymerizable MeG esters were synthesized by acylating α-MeG (3) or a protected form of MeG, methyl 4,6-O-benzylidene-α-D-glucopyranoside (8), with acrylic and methacrylic reagents, or alkylated to allyl ethers to 11 or 12 (Figure 1). Vinyl ethers of MeG were not attempted as these derivatives have already been extensively investigated (3). The results of the syntheses of polymerizable MeG derivatives are listed in Table II. Methyl 4,6-O-isopropylidene-α-D-glucopyranoside was also used as a starting material, although esters from it tended to be highly labile.

Of particular interest are the reactions of methyl 4,6-O-benzylidene-α-D-glucopyranoside (8) with acrylic and methacrylic anhydride (Figure 2). Attempts to produce diacrylates (11a) or dimethacrylates (11b) using a literature procedure (4) failed, giving only mixtures of the respective isomeric monoesters (9, 10) unless the acylation catalyst 4-

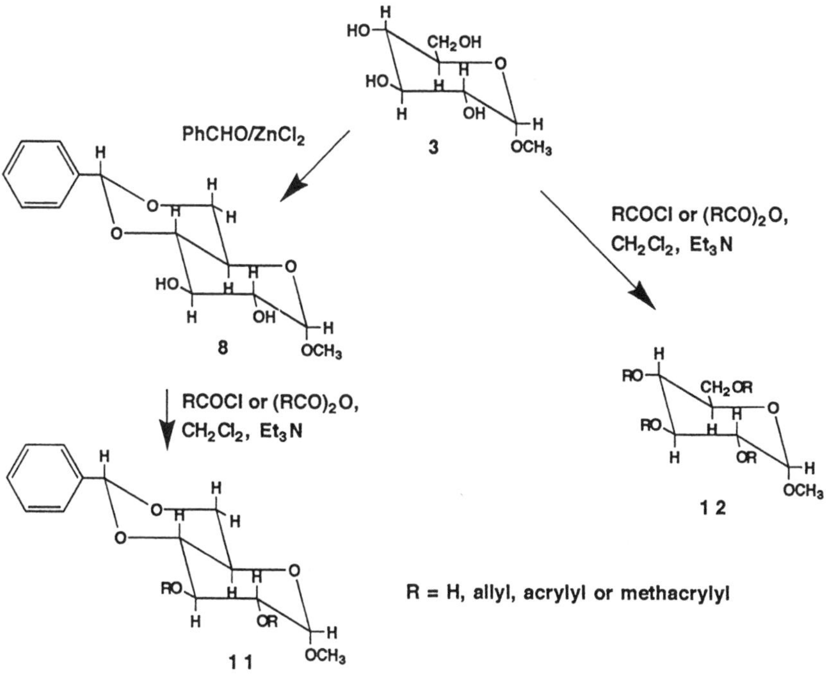

Figure 1. Synthesis of Polymerizable Methyl Glucoside Derivatives.

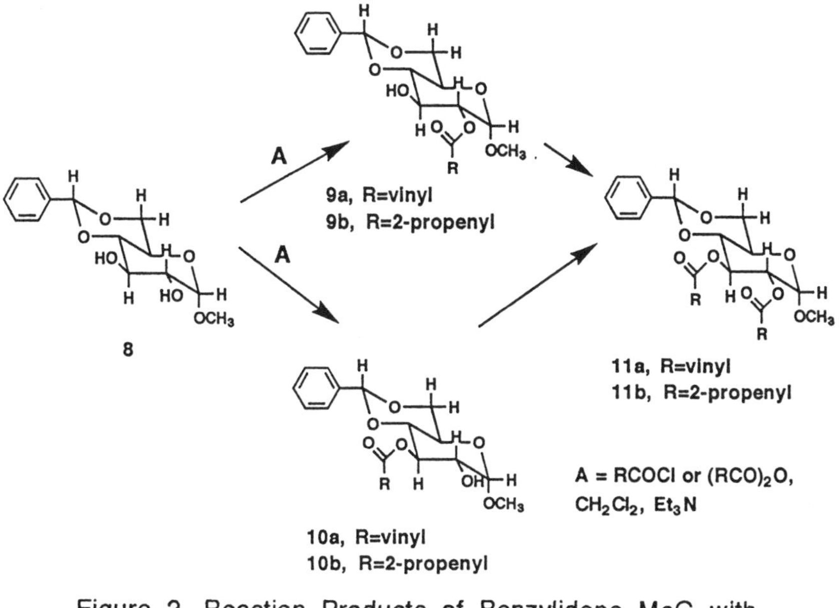

Figure 2. Reaction Products of Benzylidene MeG with Acrylic/Methacrylic Anhydrides.

Table II. Synthesis of Polymerizable MeG Derivatives

Product	Eq. Reagent/ Carbohydrate	Catalyst	D.S.[a]	% Yield[b]
Benzylidene MeG methacrylate (**9 b,1 0 b**)	1.0	none	1.3	94.0
Benzylidene MeG dimethacrylate (**1 1 b**)	2.0	DMAP	2.0	99.0
Benzylidene MeG acrylate (**9 a,1 0 a**)	1.0	none	1.0	95.1
Benzylidene MeG diacrylate (**1 1 a**)	2.0	DMAP	2.0	94.0
Isopropylidene MeG methacrylate	2.0[c]	none	1.4	99.0
Isopropylidene MeG dimethacrylate	1.0[c]	none	1.7	99.0
MeG triacrylate	3.0	none	3.0	88.6
MeG tetraacrylate	4.0	DMAP	3.9	87.0
MeG dimethacrylate	2.0	none	2.3	86.8
MeG trimethacrylate	3.0	DMAP	2.8	83.0
MeG tetramethacrylate	4.0	DMAP	4.0	99.0
Tetraallyl MeG	4.0	none	4.0	72.8

[a]Degree of substitution as determined by proton NMR
[b]Isolated
[c]Reagent eq. are correct; some decomposition occurred during work-up of each reaction to give spurious results

(dimethylamino)pyridine (DMAP) was used (*5*).

The relative amounts of the isomeric monoesters (**9, 1 0**) resembled the relative amounts of norbornyl products obtained from the Diels-Alder reaction of cyclopentadiene (**1 3**) with methyl acrylate (**1 4 a**) and methacrylate (**1 4 b**)(Figure 3)(*6*). In the Diels-Alder reaction, the more stable **1 6 b** is produced in excess when the more sterically demanding **1 4 b** is the dienophile, while the less stable **1 5 a** is the major product when the smaller **1 4 a** is the dienophile. The less stable kinetic product obtained when methyl acrylate is used is thought to arise from the conservation of unsaturation on one side of the transition state leading to **1 4 a**, which is expressed as the Alder Rule (*6*).

With 95% 2-O-methacrylate (**9 b**)/5% 3-O-methacrylate (**10 b**) produced when **8** is reacted with the sterically demanding methacrylic anhydride, and 60% 3-O-acrylate (**10 a**)/2-O-acrylate (**9 a**) from the reaction of **8** with the smaller acrylic anhydride, it appears that an Alder Rule-like interaction may be responsible for the regioselectivity of the acrylic anhydride acylation.

After the synthesis of the polymerizable MeG derivatives, a number of polymerizable derivatives of glucose, lactose, maltose and sucrose were synthesized   Allyl glucoside was among the derivatives synthesized by Fischer glycosidation, with an α/β ratio of about 1:2, from NMR analysis.   Allyl maltoside could not be made by the same procedure, as alcoholysis of the glycosidic linkage occurred to produce an identical allyl glucoside product as from glucose.   Fischer glycosidation of lactose similarly produced a mixture of allyl galactoside and allyl glucoside.

Among the other derivatives, both di- and pentamethacrylate esters of glucose and sucrose were synthesized by acylation with methacrylic anhydride.   Generally, the primary hydroxyls of sucrose were acylated, with the remaining sites in the pentamethacrylate being somewhat variable.   With glucose, the diester produced was a 1:1 α/β mixture of mainly 1,6-di-O-methacrylate.   The glucose pentaester produced by acylation was also a 1:1 α/β mixture.

Some di- and pentaallyl ethers were also synthesized from glucose by alkylation using sodium hydride/dimethyl formamide (DMF)/allyl chloride.   The diallyl ethers of glucose were mainly 1,6-substituted, about 80% β/20% α.   The pentaallyl glucose ethers were nearly 90% β-anomer.

Applications of Cornstarch-Derived MFMs.   The MeG polymerizable derivatives were generally co-polymerized with methyl methacrylate and compared to a variety of commercially available MFMs.   The co-polymerizates were tested for glass transition temperature (Tg), Izod impact and torsional modulus. Examples of some of the results are in Table III and Figure 4.

Table III, shows that many of the MeG-based additives perform as well as or better than the commercial additives in terms of increasing Tg lowering Izod impact (increasing brittleness/hardness).   In particular, the benzylidene-protected MeG derivatives performed so well that they were able to be used at one-fifth the level of the other additives and still give

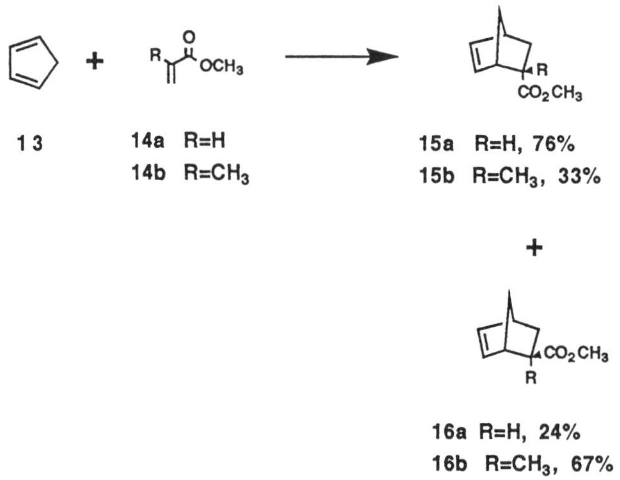

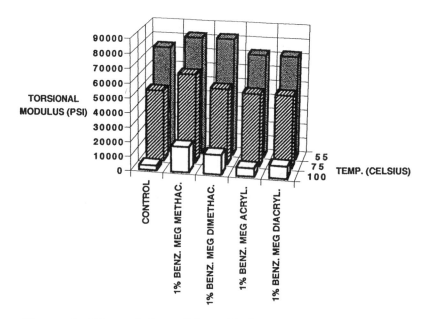

13   14a R=H   15a R=H, 76%
            14b R=CH₃   15b R=CH₃, 33%

16a R=H, 24%
16b R=CH₃, 67%

Figure 3. Examples of Diels-Alder Reaction.

Figure 4. Effect of Benzylidene MeG Additives on Torsional
Modulus of Poly(Methyl Methacrylate).

Table III. Effect of Commercial and MeG-Based Additives on Properties of Poly(Methyl Methacrylate)

Additive	Poly. Groups/ Molecule	Wt. % Added	Tg (°C)	Izod Impact[a]	Gardner Color[b]
None (control)	-	-	100	C3=0.37	1
Commercial Additives					
Glycerolpropoxy triacrylate	2.7	5.0	115	C2=0.20	1
Pentaerythritol triacrylate	3.3	5.0	125	C3=0.27	1
Pentaerythritol tetraacrylate	4.0	5.0	111	C3=0.19	1
Pentaerythritol tetramethacrylate	4.0	5.0	117	C2=0.28	1
Trimethyolpropane triacrylate	2.7	5.0	118	C1=0.19	1
Trimethyolpropane trimethacrylate	2.7	5.0	102	C3=0.15	1
MeG Additives					
Benz. MeG methac.	1.3	1.0	103	C3=0.24	1
Benz. MeG dimethac.	2.0	1.0	108	C2=0.26	1
Benz. MeG acrylate	1.0	1.0	108	C3=0.28	1
Benz. MeG diacrylate	2.0	1.0	114	C3=0.29	1
Isop. MeG methac.	1.4	5.0	105	C1=0.21	1
Isop. MeG dimethac.	1.7	5.0	104	C2=0.20	1
MeG triacrylate	3.0	5.0	112	C1=0.21	1
MeG tetraacrylate	3.9	5.0	107	C1=0.19	3
MeG dimethacrylate	2.3	5.0	111	C3=0.21	6
MeG trimethacrylate	3.0	5.0	81	C2=0.20	4
MeG tetramethac.	4.0	5.0	116	C3=0.26	4
Tetraallyl MeG	4.0	5.0	104	C3=0.19	1

[a]Letter indicates type of break, C(n)=clean break, n samples
[b]Higher number indicates darker color; 1=transparent

comparable results.

Figure 4 shows the effect of the benzylidene MeG additives on torsional modulus, a measure of the amount of force required to twist a sample with changing temperature. As can be seen, the additives typically causes an increase in the amount of force required at most temperatures, especially 100°C. Although not shown, the torsional modulus results for the MeG additives were generally equal or superior to the results obtained with the commercial additives.

The glucose/sucrose polymerizable derivatives were co-polymerized with methyl methacrylate to directly compare to the MeG derivatives, and were also graft co-polymerized with poly(propylene). The analysis of the poly(methyl methacrylate) co-polymerizates is still in progress; the graft co-polymers with poly(propylene) were tested for stress, the load applied to a cross-sectional area of a polymer, and modulus of elasticity, the force needed to elongate a polymer, using a three-point bend test.

Examples of some of the results are in Figures 5 and 6. It is apparent in Figure 5 that the grafting process alone, in which a peroxide catalyst is added to poly(propylene), somewhat increases the stress, but not nearly as much as the dimethacrylyl glucose. The superiority of dimethacrylyl glucose in increasing the modulus of elasticity over the allyl derivatives is evident in Figure 6. In general, the methacrylate derivatives appeared to be better cross-linkers than the allyl ether derivatives, although some of the allyl glycosides performed nearly as well as the methacrylates.

Thermoset Applications

Phenol-Formaldehyde Resin Reactions. There are two types of polymers produced from the reaction of phenol and formaldehyde: 1) novolak or acid-catalyzed resins and 2) resol or base-catalyzed resins. They differ, other than in pH, primarily in that Novolak resins have more ether linkages than resols. It is the resol resins that are of interest for adhesives for gluing wood or that will come in contact with wood, as the acid-catalyzed novolaks degrade cellulose.

Fiberglass Insulation Binders. Fiberglass insulation bats used in walls and ceilings are bound together by resol PF resins. These resins have a formaldehyde/phenol (F/P) ratio of about 4.0, much higher than the 2.1 F/P of wood adhesive resins, because the higher F/P allows them to cure much faster. Also, extenders, such as lignin, are often used.

The effect of methyl glucoside on insulation properties are summarized in Table IV. Most notable is the lack of flame spread.

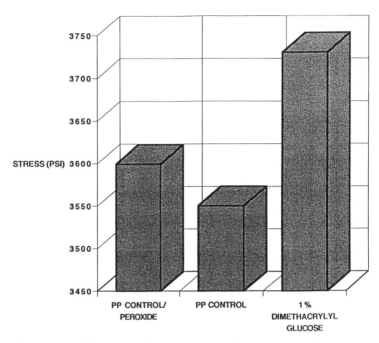

Figure 5. Effect of Dimethacrylyl Glucose on Stress of Poly(Propylene) Graft Co-Polymers.

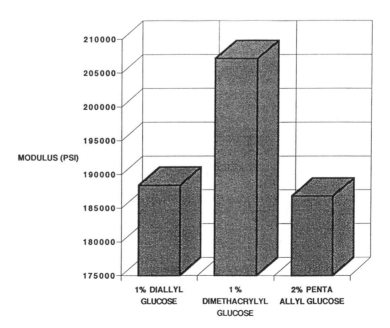

Figure 6. Effect of Glucose Additives on Modulus of Elasticity of Poly(Propylene) Graft Co-Polymers.

Table IV. Properties of Insulation Binders Containing MeG[a]

% MeG	% Lignin	Height Recovery (inches)	Dry Parting Strength (ft./min.)	Flame Spread (g/g)	Moisture Absorption (%)
0	0	6.59	155	7	0.93
0	16	6.99	130	5	1.49
16	0	7.00	154	0	0.96
30	0	6.84	117	0	1.65

[a]Mixture of methyl glucosides

Emissions from the binder are carefully monitored, of course, as these emissions could affect people in their homes, the most common site of fiberglass insulation use. Table V summarizes the effect of MeG on binder emissions.

Table V. Emissions From Insulation Binders Containing MeG[a]

% MeG	% Lignin	Amines (ppm)	Formaldehyde (ppm)
0	0	8.2	2.6
0	16	6.3	8.1
16	0	4.3	6.4
30	0	2.7	5.0

[a]Mixture of methyl glucosides

Plywood Adhesives. The idea of incorporating carbohydrates into phenol-formaldehyde (PF) resin-based adhesives is, or course, not new. As early as 1926, carbohydrates were being incorporated into PF resins (7). A number of studies have followed, including the classic work of Chang and Kononenko, who used sucrose (8). More recently, a number of studies have come out of the USDA Forest Products Laboratory (9), including one where a variety of carbohydrates and other polyols were tested in plywood glues (10). However, even with all this history and experimentation, mixtures of MeG anomers are currently the only carbohydrates being used to extend or replace PF resins in plywood glues and insulation binders (11).

One reason for the lack of utilization of carbohydrates in PF resins is a lack of understanding of their reactivity with resin components. Early investigations involving the incorporation of reducing sugars into PF resins assumed dehydration of the sugar to furfural derivatives if the pH of the reaction was neutral or less (*12*). However, in the base-catalyzed Resol resins that are normally used as wood adhesives, no furan resonances are observed by carbon-13 nuclear magnetic resonance (NMR) when reducing sugars were incorporated (*9a, 13*). Other investigators have stated that reducing sugars cannot be used under basic conditions because of base-catalyzed degradation reactions that produce saccharinic acids which neutralize the basic catalyst (*14*). Finally, a recent report claims that the hydroxyls of non-reducing carbohydrates react with phenolic methylol groups to form ether linkages (*15*).

MeG, a non-reducing sugar, fits into this last category. As there was no definitive method as to the best mode of incorporating a MeG mix into PF resins, attempts were made to incorporate the MeG mix into both resin cooks and glue mixes (*16*) Some attempt was also made to determine MeG reactivity, if any, with resin components. Fillers in plywood adhesives are defined as being unreactive, while extenders are defined as having reactivity, so the reactivity of MeG with resin components is important in terms of how to classify MeG.

Plywood Glues. Initial attempts to incorporate the MeG mix into plywood adhesives concentrated on addition to glue mixes. Two approaches were followed: simple addition to glue mixes and partial replacement of PF resin. Laboratory studies indicated several positive properties, in addition to lower cost, caused by the presence of MeG, especially dry-out resistance and better flow properties of the glue. The dry-out resistance imparted by MeG to the glue mix allowed significant spread reductions, resulting in further cost savings. MeG mixtures were tested in large-scale trials at several southern yellow pine plywood mills, with the results given in Table VI, showing little change in wood failure of panels made with or without MeG for two plywood mills over 6-12 months. The wood failure results for panels made with MeG in the glue mix were achieved with a 7-12% spread reduction, significantly reducing costs. Details about the incorporation of MeG into glue mixes were recently published (*17*).

Table VI. Wood Failure Results from Two Plywood Mills Using MeG[a] In Glue Mix[b]

Result	Mill 1		Mill 2	
	Without	With	Without	With
Wood failure average[c] (%)	87.5	86.5	89.1	89.1
Panels with averages below 60% wood failure (% of total panels)	4.1	4.0	3.2	2.9

[a]Mixed product

[b]Adapted from Sellers, T. Jr. and W.A. Bomball, *Forest Prod. J.*, 1990, **4 0**, 52.

[c]Wood failure averages are based on independent testing agency estimations made on test specimens subjected to tension shear after a vacuum-pressure soak test.

Resin Cooks. A simpler, more direct incorporation of mixtures of MeG anomers into plywood adhesives can be achieved by cooking the MeG mix into PF resins, which can then be distributed to numerous plywood mills, as opposed to attempting to optimize the MeG mix in every plywood mill's glue mix. As MeG demonstrated dry-out resistance in glues, it was thought that simply cooking a MeG mixture into a resin might allow production of a higher molecular weight resin--in other words, a single cook resin with double cook properties (*18*). It was hoped that MeG might also allow a much higher F/P ratio than normal for plywood resins without making the resin viscosity unmanageable or sacrificing storage stability (*10*).

Initial plans for cooking a MeG mix into PF resins called for making a series of single cook resins with F/P ratios up to 2.7 and up to 20% of the resin replaced by MeG (see the dotted-line box in Figure 7). In order to maintain constant resin solids, both phenol and formaldehyde were replaced by MeG versus an unmodified resin of identical F/P. Figure 8 shows the wood failure results for test panels made from these MeG-modified resins on southern yellow pine veneer. Clearly, there is a trend toward increasing % wood failure with increasing F/P and % MeG in the resin.

It should be noted that while the wood failure numbers might appear to be lower than industry acceptable numbers (80% or greater wood failure), these numbers are low because of the

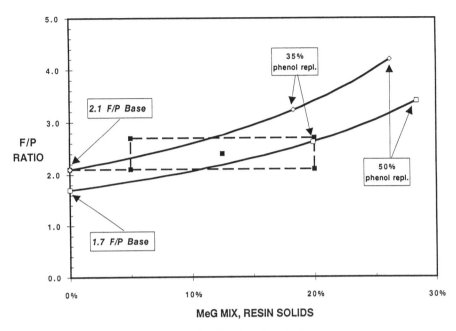

Figure 7. Resin Synthesis Plan.

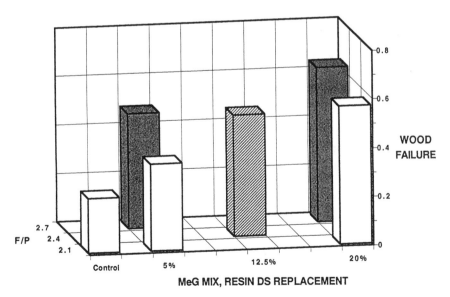

Figure 8 Effect of MeG Cooked in PF Resins on Wood Failure.

experimental design. The parameters used to produce the test panels are intentionally at the fringe of normal operating parameters to magnify differences in the resins. Therefore, minimum spreads (28-33 lbs./MSGL) and hot press times (3 min.) and high veneer moisture contents (8-9%) were used to stress the system, conditions that with normal resins might be expected to produce low wood failures. Under optimum conditions (higher spreads, longer hot press times), nearly all resins produce satisfactory (80% or greater wood failure) test panels, but such conditions do not allow differentiation between resins. Under the stressed conditions, the MeG-modified resins produced significantly better results than unmodified control resins.

Since the initial results indicated that better results might be obtained with higher F/P ratios and MeG mix levels, resins with both of these characteristics were synthesized as shown shown by the curves in Figure 7. This time, instead of replacing both phenol and formaldehyde in the resin with MeG to maintain solids levels for each F/P ratio, only phenol was replaced by MeG, resulting in a doubling of the F/P if 50% of the phenol in a 2.1 F/P resin was replaced by MeG. Cook profiles (not shown) of 35% phenol replacement resin (3.23 F/P) compared to a 2.1 F/P control indicated that both resins could be cooked to identical viscosities in identical 4 hour cook times, but the MeG-modified resin required approximately 5-7 degrees higher temperature (10 degrees higher for 50% phenol replacement resins, 4.2 F/P) to achieve this cook time. Once again, these are single cook resins.

Figure 9 shows % wood failure for the phenol replacement resins versus controls as a function of hot press time and starting F/P ratio. The % wood failure increases for all the resins with increasing hot press time, but the increase is much larger for the control resin, indicating that the phenol replacement resins can function much better at marginal hot press times. There is little difference in the performance of the phenol replacement resins versus controls as a function of the starting F/P ratio. In the case of a starting F/P ratio of 1.7, a maximum F/P of 3.4 was achieved when 50% of the phenol was replaced, as opposed to an F/P of 4.2 when replacing 50% of the phenol in a 2.1 F/P resin. This startling performance was attained using 28 lbs. MSGL, 4 min. hot press time and 9% veneer moisture.

Since the control resins performed well versus the more inexpensive, high F/P phenol replacement resins at longer hot press times, high F/P controls were synthesized to compare their performance to the modified resins at short hot press times. However, as the F/P ratio was increased to 2.7 without the presence of MeG, free formaldehyde levels in the controls increased to over 3% (see Figure 10). The gel times for these controls also decreased by about 75%, indicating little storage stability. Because of the strong formaldehyde fumes, and dry-out at longer assembly times, these resins were unusable.

Additional resins synthesized included resins with F/P ratios of 3.2, each containing approximately 20% sucrose and 20% pure methyl α-D-glucopyranoside. Both of these resins were unusable due to free formaldehyde in excess of 3% and gel times 75% less than the high F/P MeG phenol replacement resins. These results seem to confirm a major difference in the reactivity of pure MeG versus the mixed product. The furanosides and higher oligosaccharides present in the mixed product apparently are a major factor in the much better performance of the resins in which they are present. Whether this is because of higher reactivity or other reasons is not yet known.

One of the most important results obtained to date in the research involving MeG in PF plywood resins was the effect of MeG on free formaldehyde. Figure 10 shows the effect of MeG on high F/P resins. Compared to 2.7 F/P control resins, resins with 4.2 F/P containing MeG had lower free formaldehyde levels, allowing them to be used in plywood applications (*19*). This scavenging of formaldehyde by MeG was also noted in fiberglass resins, as is an important feature of the performance of MeG in other applications, such as textile finishing resins.

Textile Finishing Resins

Methyl glucoside has been used in textile finishing, both as a weighter for knit fabrics and as a formaldehyde scavenger in durable press finishing resins. As these applications are extremely color sensitive, with pHs on the acidic side, applications of reducing sugars are limited, and a pure form of MeG, with little or no residual reducing glucose, namely methyl α-D-glucopyranoside, must be used. The results are listed in Table VII.

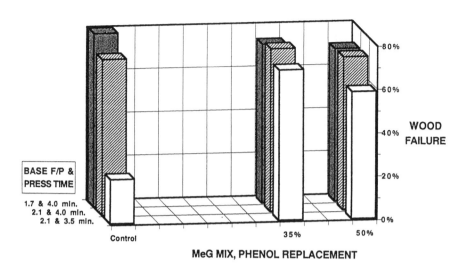

Figure 9. Effect of MeG Replacing Phenol in High F/P PF Resins on Wood Failure as a Function of Base F/P and Press Time.

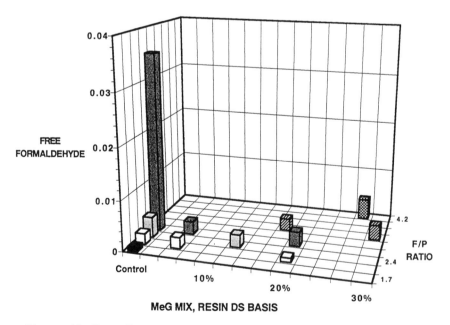

Figure 10. Free Formaldehyde in PF Resins as Function of MeG Level and F/P Ratio.

Table VII. Formaldehyde Release Using 4% MeG[a] in Finishing Resin on Knit Fabrics[b]

Resin	Formaldehyde (ppm)[c]	Reduction (%)
Control 1	370	
Control 1 + 4% MeG	190	48.6
Control 2	415	
Control 2 + 4% MeG	185	55.4
Control 3	430	
Control 3 + 4% MeG	290	32.6
Control 4	400	
Control 4 + 4% MeG	225	43.8

[a]Pure methyl α-D-glucopyranoside

[b]Work done at North Carolina St. University School of Textiles for A.E. Staley Mfg. Co. under the direction of Professor C. Tomasino

[c]Determined by American Association of Textile Colorists and Chemists test method 112-1984.

Other Applications

Cornstarch-derived chemicals like MeG have been utilized in other applications, such as polyurethane foams (*20*). Additional work in this area has been done at the National Center for Agricultural Utilization Research (*21*).

Future Considerations

With the ability to scavenge formaldehyde, cornstarch-derived carbohydrates may have applications wherever formaldehyde is used. In the area of wood adhesives, with only plywood having been extensively researched, there are numerous other types of structural panels in which cornstarch-derived chemicals may be usable. These include fiberboard, oriented strandboard (OSB), particleboard and waferboard. All of these panel types use lower quality wood and more adhesive than plywood, so the potential markets are even greater.

Another application that uses formaldehyde and where cornstarch-derived chemicals may have use is in melamine-formaldehyde (MF) laminating resins. Decorative laminates, such as countertops and panelling, and industrial laminates, such as circuit boards, are applications of MF laminating resins.

Polyurethanes are another possible area of further application of cornstarch-derived chemicals. As mentioned above, some work has already been done with MeG; further work in this area has been done with other cornstarch-based materials, with more research needed to commercialize more carbohydrate materials in this area.

Summary and Conclusions

As seen from the data presented here, cornstarch-derived chemicals like glucose and MeG have numerous applications. In many cases, the properties derived from such materials are equal or superior to commercial petroleum-derived chemicals. But with lower cost, stable source of supply and the ability to scavenge formaldehyde, cornstarch-derived chemicals have some unique advantages. Yet, at this writing, they still have not been commercialized to any great extent.

Industry is very slow to change from well-known petroleum-based materials with a long history, even given the advantages of the cornstarch-derived chemicals. However, with increasing regulatory pressure concerning health and safety of consumers and workers, and with higher oil prices possibly in the future, cornstarch-derived chemicals will eventually find larger and more receptive markets.

Acknowledgements

The following sources of equipment and funds for much of the research described in this chapter are gratefully acknowledged: Grain Processing
Corp., Muscatine, IA; Illinois Corn Marketing Board, A.E. Staley Mfg. Co., Decatur, IL; United States Department of Agriculture/Illinois Agricultural Experiment Station; State of Illinois Value-Added Program; University of Illinois at Urbana-Champaign/Campus Research Board and University of Illinois at Urbana-Champaign/Department of Food Science.

Not to be forgotten are former co-workers at A.E. Staley Mfg. Co., who participated in much of the research described herein: Raymond L. Drury, Jr., Dr. George B. Poppe and Patricia S. Tippit. And last, but not least, my current co-workers at the University of Illinois at Urbana-Champaign/Department of Food Science: graduate students Larry P. Karcher, who has worked on thermoset applications both at Staley and at the university, and William J. Krueger, who did much of the thermoplastic applications work.

Literature Cited

1. From "The World of Corn," National Corn Grower's Association, 1990, pp. 8, 20-21, 28.
2. See, for example, numerous articles in Proceedings of Corn Utilization Conference II, Corn Growers Association, Columbus, OH, 1988, Proceedings of Corn Utilization Conference III, Corn Growers Association,.St. Louis, MO, 1990 and Proceedings of Corn Utilization Conference IV, Corn Growers Association, St. Louis, MO, 1992.
3. Marvel, J.T., J.W. Berry, R.O. Kuehl and A.J. Deutschman, *Carbohyd. Res.*, 1969, **9**, 295.
4. Haworth, W.N., H. Gregory and L.F. Wiggins, *J. Chem. Soc.*, **1946**, 488.
5. Dunn, L.B. Jr., U.S. Patent 4,833,202, May 23, 1989.
6. Carey, F.A. and Sundberg, "Advanced Organic Chemistry," Part A, Plenum Publishing Co., New York,
7. Meigs, J.V. U.S. Patent 1,593,342, 1926.
8. Chang, C.D. and O.K. Kononenko, *Adhesives Age*, 1962, **5**, 36.
9. a. Christiansen, A.W., Proceedings of Adhesives Conference, Forest Products Laboratory, Madison, WI, 1985, p. 211. b. Chrisitiansen, A.W. and R.H. Gillespie, *Forest Prod. J.*, 1986, **36**, 20 and references therein.
10. Conner, A.H., B.H. River and L.F. Lorenz, Proceedings of Adhesive Conference, Forest Products Laboratory, Madison, WI, 1985, p. 227.
11. Personal communication with A.E. Staley Mfg. Co., Decatur, IL.
12. Gibbons, J.P. and L. Wondolowski, Canadian Patent 1,090,026, 1980.
13. Clark, R.J., J.J. Karchesy and R.L. Krahmer, *Forest Prod. J.*, 1988, **38**, 71.
14. Conner, A.H., B.H. River and L.F. Lorenz, *J. Wood Chem. Tech.*, 1986, **6**, 591.
15. Ibid, "Carbohydrate-Modified Phenol-Formaldehyde Resins Formulated at Neutral Conditions," in "Adhesives From Renewable Resources," R.W. Hemingway and A.H. Conner, eds., American Chemical Society, Washington, D.C., 1988, chapter 25.

16. Glue mixes are made at each plywood mill immediately   prior to plywood manufacture.

17. Sellers, T. Jr. and W.A. Bomball, *Forest Prod. J.*, 1990, **4 0**, 52.

18. Avery, L.P., W.A. Bomball, R.L. Drury, Jr., L.B. Dunn, Jr., B.J. Hipple and M.C. Kintzley, U.S. Statutory Invention Registration No. H603, 1989.

19. The effect of MeG on PF and urea-formaldehyde (UF) wood adhesive resins was recently presented at a Forest Products Research Society National Meeting.

20. See STA-MEG^{TM} 104 Methyl Glucoside Technical Brochure, A.E. Staley Mfg. Co., Decatur, IL   62525, 1984.

21. See, for example, Cunningham, R.L., M.E. Carr, E.B. Bagley and T.C. Nelson, *Starch*, 1992, **4 4**, 141 and Cunningham, R.L., M.E. Carr and E.B. Bagley, *J. Appl. Polym. Sci.*, 1992, **4 4**, 1477.

RECEIVED May 24, 1994

Chapter 10

Simple Preparation of Hydroxylated Nylons—Polyamides Derived from Aldaric Acids

D. E. Kiely[1], L. Chen[1], and T-H. Lin[2]

[1]Department of Chemistry, University of Alabama at Birmingham, Birmingham, AL 35294–1240
[2]Avanti Polar Lipids Inc., 700 Industrial Park Drive, Alabaster, AL 35007

The large scale availability of D-glucose (dextrose) from starch hydrolysis makes this sugar an attractive starting material for industrial scale polymer syntheses. With that prospect in mind we developed a simple and versatile process for the preparation of "Hydroxylated Nylons" (polyhydroxypolyamides) from D-glucaric acid, the aldaric acid from D-glucose oxidation. A number of such polyamides have been prepared by our newly modified process, which utilizes the monopotassium salt of D-glucaric acid as the starting material for the carbohydrate diacid monomer. The process, representative polymers prepared using the process, and some notable property differences within this family of polyamides are described.

Much of our laboratory research involves the use of unprotected carbohydrate molecules in organic synthesis. This research is motivated by a need for development of new, large scale industrial processes that utilize monosaccharides from hydrolysis of oligosaccharides or polysaccharides which are available from agricultural sources. However, for such processes to be industrially practical and economical, the use of carbohydrate protection/deprotection steps, a common feature of laboratory scale carbohydrate syntheses, must be avoided. The monosaccharides of greatest interest to us are D-glucose (dextrose), D-xylose and D-galactose. These monosaccharides are hydrolysis products of primarily corn starch, wood xylans, and milk lactose, respectively. D-glucose is clearly at the top of the list of important monosaccharides because billions of pounds of "liquid dextrose" are available each year from corn starch hydrolysis, particularly in the United States. Yet, dextrose still remains a largely untapped chemical reservoir for novel, major scale industrial synthetic processes for generating new and useful products, products with the potential added benefit of good biodegradability characteristics. This paper briefly describes some of our efforts to develop chemistry of unprotected D-glucose derivatives that might be useful on an industrial scale to generate a family of new and structurally variable carbohydrate polymers which we refer to as "Hydroxylated Nylons".

0097–6156/94/0575–0149$08.00/0

Can Strictly Synthetic Polymers be Made Conveniently From Carbohydrates Using Standard Polymer Technology?

An increased awareness among scientists and the general public in recent years of the availability of useful and biodegradable disposable items and products has spurred interest in development of biodegradable polymers from carbohydrate derived starting materials. Perhaps most notable among such polymers is poly(lactic acid) (*1*), derived from the condensation of L-lactic acid produced from a D-glucose fermentation process (*2*).

Our interests are in the preparation of synthetic copolymers using step-growth or condensation polymerization methods common to the production of high volume synthetic polymers such as nylons (*e.g.*, nylon 6,6) and polyesters [*e.g.*, poly(ethylene terephthalate)]. A proposed approach for the general preparation of synthetic carbohydrate based copolymers parallels that used for strictly petroleum based monomers, and is shown in Scheme 1. A key structural feature of the carbohydrate monomer is that the **carbohydrate monomer is in an unprotected acyclic form activated at both ends of the monomer unit.**

Scheme 1. Synthetic Copolymers Derived from Carbohydrates

$$-A-B-A-B-A-B-A-B-A-B-A-B- \qquad i.e., \; -[A-B]_n-$$

Polymerization Reaction - Condensation or Step-Growth Polymerization

$$nA + nB \longrightarrow -[A-B]_n-$$

Monomer A - derived from $X-(CHOH)_Y-X$, an acyclic unprotected carbohydrate activated at both ends.

Monomer B - derived from $Z— Z$, a diterminally activated second monomer.

Some Critical Characteristics of Idealized Carbohydrate Based Polymerizations For Industrial Production.

Because basic research of this type is ultimately directed at industrial scale production of new carbohydrate polymers, we have had to consider some characteristics of idealized carbohydrate based polymerizations that are essential for large scale production.

1. Activated carbohydrate monomers **A** are available from any simple aldose; *e.g.*, D-glucose, D-galactose, D-xylose, D-mannose, etc.
2. Monomers **B** are relatively inexpensive and readily available.
3. Syntheses of the activated carbohydrate monomers **A** and the polymers $-[A-B]_n-$ do not require protection-deprotection steps.

4. Polymer isolation and purification utilizes established industrial technology.
5. Polymers with reasonably predictable but variable properties can be synthesized by taking into account:
   a. the structure of the carbohydrate monomer **A** (*i.e.*, stereochemistry and number of carbons in the carbohydrate;
   b. the chain length, degree of branching, and funtionalization of monomer **B**.

"Hydroxylated Nylons" - Target Carbohydrate Based Polymers.

With the above polymerization characteristics in mind, our target molecules were carbohydrate based "Hydroxylated Nylons". Such target molecules are illustrated in **Figure 1** with poly(hexamethylene D-glucaramide) as a specific example. Also shown is its strictly petroleum based structural counterpart, commercial nylon 6,6. Nylon 6,6 is produced under high temperature conditions from an adipic acid - hexamethylenediamine salt, a process that drives off water in formation of amide bonds. Such conditions are far too severe for making polyamides from carbohydrate diacids (aldaric acids) due to unwanted dehydrations and other degradative reactions that the carbohydrate monomers and polymers might undergo. Consequently, much lower temperature conditions for polymerization from D-glucaric acid had to be developed.

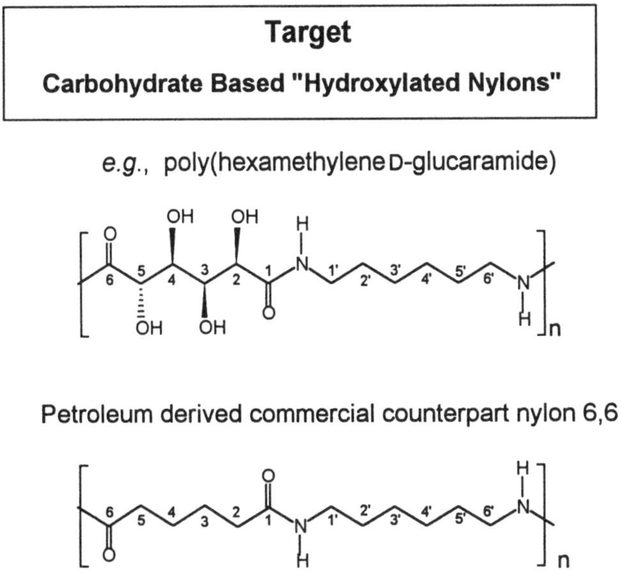

Target

Carbohydrate Based "Hydroxylated Nylons"

e.g., poly(hexamethylene D-glucaramide)

Petroleum derived commercial counterpart nylon 6,6

Figure 1. Target Carbohydrate Based Hydroxylated Nylon

General Scheme for Preparing Poly(alkylene D-glucaramides).

Scheme 2 illustrates the general process we are presently using to prepare poly(alkylene D-glucaramides) from D-glucose(3-4). D-glucose as starting material for the carbohydrate monomer, is oxidized with aqueous nitric acid based upon procedures dating back more than 100 years. D-Glucaric acid is then conveniently isolated as its crystalline monopotassium salt which in turn is directly converted to a mixture of ester/lactone forms (labelled as Activated D-Glucaric Acid). This mixture can be used directly, without separation of the various ester forms of glucaric acid in the mixture, for condensation with a diamine(s) of choice. Condensation is generally done in a polar solvent such as methanol, at room temperature to produce the polyamide. Most often the polyamide precipitates in solid or amorphous form from the solvent and after washing with solvent requires no purification. The triethylamine in the reaction mixture catalyses the reaction to completion at an appreciable rate.

Scheme 2. General Scheme for Preparing Poly(alkylene D-glucaramides)

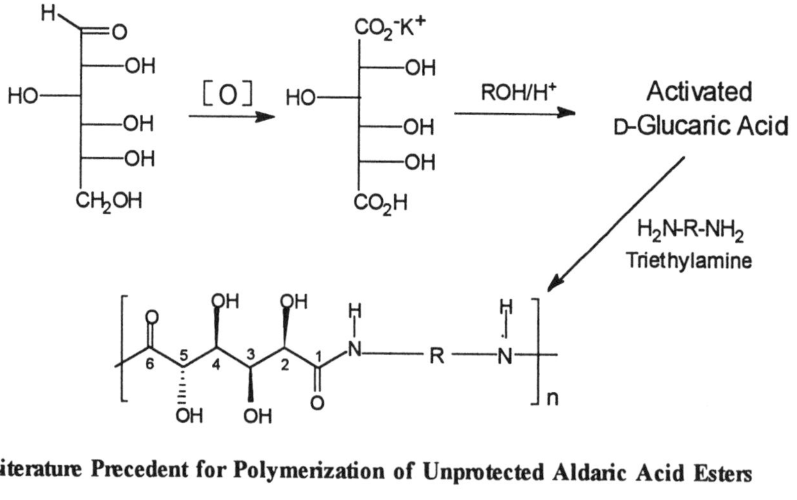

Literature Precedent for Polymerization of Unprotected Aldaric Acid Esters

While the process in Scheme 2 illustrates the conversion of D-glucaric acid to Hydroxylated Nylons, Ogata and coworkers in the 1970's (5-11) pioneered the preparation of similar polyamides from condensation of diethyl mucate (diethyl galactarate) and dimethyl tartarate with diamines, including hexamethylenediamine. Those condensations were carried out in polar solvents that included methanol, methyl sulfoxide and N-methylpyrrolidone. Those workers noted diesters with heteroatoms such as oxygen and sulfur bonded to the carbon α to ester carbonyl carbons underwent polymerization with diamines under very mild conditions. Hoagland significantly clarified the mechanism of the carbohydrate diester/amine condensation in his studies of the aminolysis of diethyl galactatrate (*12*) and diethyl xylarate (*13*). The Hoagland mechanism is a two step sequence; a fast base catalyzed five-membered lactonization step followed by a slower lactone aminolysis step. Hoagland's results clearly point out that activation of five and six carbon aldaric acid diesters is a consequence of the ease of formation and high reactivity of the *in situ* generated five-membered aldarolactones under the basic conditions of condensation with amines.

Activated Glucaric Acid as a Condensation Polymerization Monomer.

The activated forms of D-glucaric acid which facilitate the polymerization of aldaric acid with diamines are an equilibrium mixture of the ester/lactones shown in **Figure 2**. When the monopotassium salt of D-glucaric acid is treated with an alcohol (*e.g.*, methanol or ethanol) solution of HCl or with an insoluble acid cation exchange resin in its H^+ form, a mixture of dialkyl D-glucarate (**1**), alkyl D-glucarate 1,4-lactone (**2**), alkyl D-glucarate 6,3-lactone (**3**) and D-glucaro-1,4:6,3-dilactone (**4**) is formed *in situ* (*14*). After acid removal, the equilibrium mixture in alcohol solution can be directly condensed with the primary diamine of choice to produce the polyamide. This method of preparation of activated glucaric acid has an energy saving advantage over an earlier reported method (*15-16*) in that conversion of the salt form to the ester forms does not require the use nor removal of solvent water.

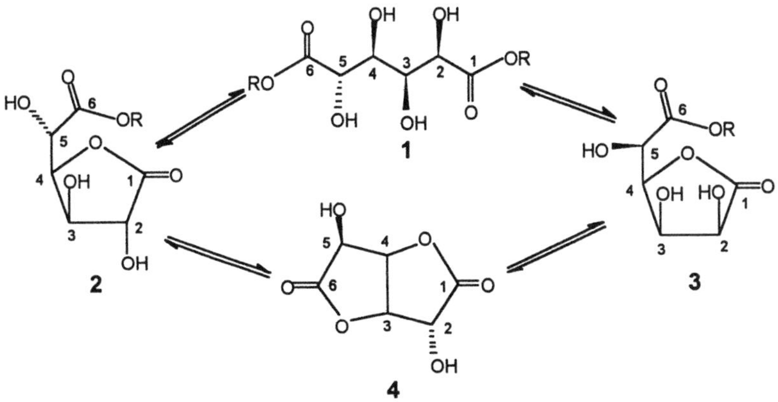

Figure 2. Equilibrium mixture of esterified D-glucaric acid in basic alcohol solution

Convenient to use and easily obtained crystalline forms of activated D-glucaric acid that allow good stoichiometric control of the diacid monomer in polymerizations are the ester lactones **2** (R = Me) or **3** (R = Et) (*14*) and the dilactone **4** (*17*). The dilatone **4** has been employed by Hashimoto and coworkers as a monomer for condensation with *p*-xylenediamine to the corresponding polyamide (*17*). However, the ester/lactones **3** & **4** are particularly useful in that they appear to have good shelf live and are not readily hydrolyzed on standing, as is the dilactone **4**.

Some Poly(alkylene D-glucaramides) Prepared Directly From Monopotassium D-glucarate; Structures and Comparison of Some Physical Properties.

Figure 3 shows the structures of some of the structurally variable polyamides that are conveniently made using a common procedure. These polyamides, related polyamides, and the experimental details for preparing the polyamides are more fully described in references *3* and *4*.

The strictly poly(alkylene D-glucarates), poly(hexamethylene D-glucaramide) and poly(tetramethylene D-glucaramide), **5** & **6**, are both high melting white solids (192-195 & 192-194 °C, respectively) but with very different solublility properties. Whereas polyamide **5** and polyamides with additional methylene units (*3-4*) apppear to have no water solubility, polymer **6** and the ethylenediamine polymer (*3-4*) are readily soluble in water at room temperature. Polyamides **7** & **8**, poly(3',6'-dioxaocta-methylene D-glucaramide) and poly(4'-aza-*N*-heptamethylene D-glucaramide) respectively, are examples of polymers with heteroatoms in the alkylene chain. Polymers **7** & **8** are also readily water soluble and have considerably lower melting points (both 150 °C) than do polymers **5** & **6**. Polymer **8** contains a tertiary amine group that can be further alkylated, such as shown with the quaternary ammonium containing salt, water soluble polymer **9**. The last polyamide shown (**10**), poly(*m*-xylylene D-glucaramide) is a high melting white solid (210-215 °C) and is water insoluble.

Some Computer Aided Structural Studies of Hydroxylated Nylons.

In order to try to better understand the observed variance in physical properties of Hydroxylated Nylons derived from D-glucaric acid (*18*), and other aldaric acids (*19*), with various diamines, we have carried out some preliminary molecular modeling studies concerned with evaluating the secondary structure of the polyamides in solution. The ^{1}H NMR spectra of many of these polyamides have been recorded, using DMSO-d_6 or trifluoroacetic acid as the NMR solvent. The similarity of the carbohydrate proton chemical shifts and vicinal coupling constants found in D-glucaric acid based polyamides *e.g.*, poly(hexamethylene D-glucaramide), PHGA] and the corresponding *N*,*N*'-dialkyl D-glucaramides (*N*,*N*'-dihexyl D-glucaramide, DHGA) suggests that the conformation of the D-glucaric acid moiety of the polyamides and diamides in solution are also similar. The appropriate ^{1}H NMR data from some gluaramides and a poly(xylaramide) are given in the Table. Employing molecular modeling (MM2) methods, the relative energies of some glucaramide (acyclic) conformations were determined (**Figure 4**) (*18*). The sickle conformation labeled *Sickle-1*, **B**, was calculated to be of lowest energy and was chosen as the conformtional D-glucaramide unit in generating computer derived drawings of the poly(D-glucaramides) shown in **Figure 5**. The vicinal coupling constant data from the ^{1}H NMR spectra of these polymers also supports the notion that *Sickle-1*, **B** is the preferred conformation of the D-glucaramide monomer unit. This sickle conformation is very reasonable in that it relieves the unfavorable steric interaction between the hydroxyl groups at C-2 & C-4 of the D-glucaramide monomer unit.

Figure 5 shows the computer generated end on and side views of three poly(glucaramides): **A**, poly(hexamethylene D-glucaramide); **B**, poly(*m*-xylylene D-

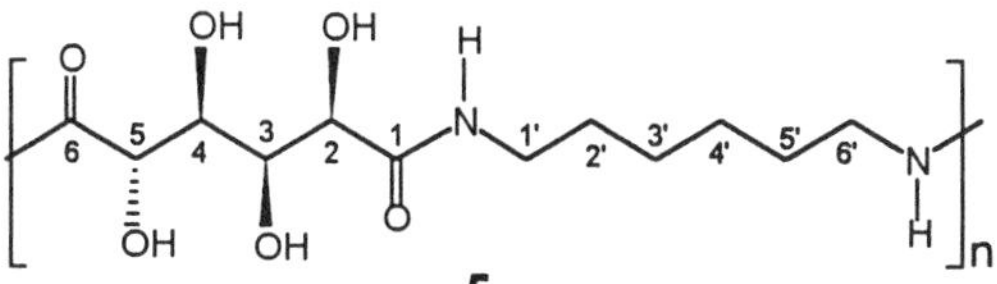

5

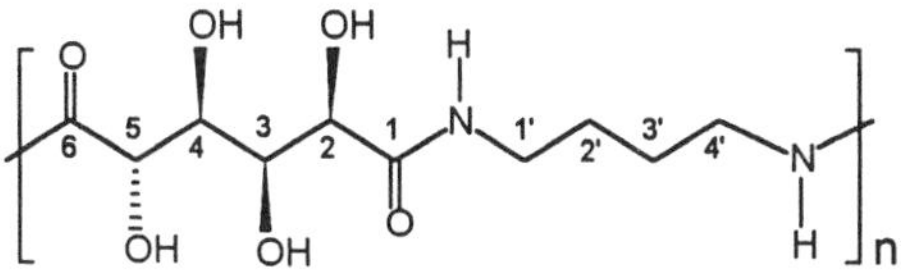

6

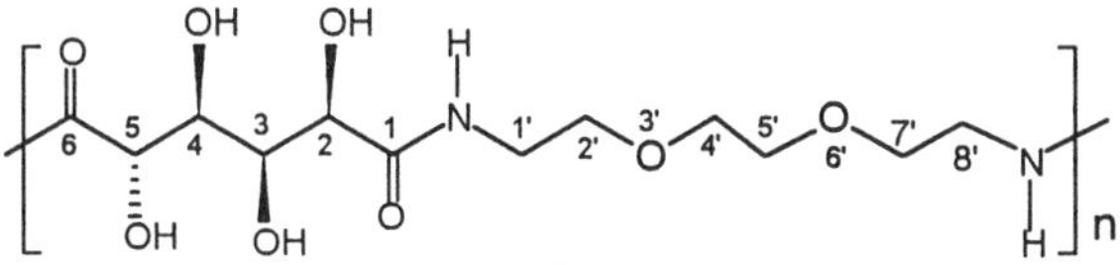

7

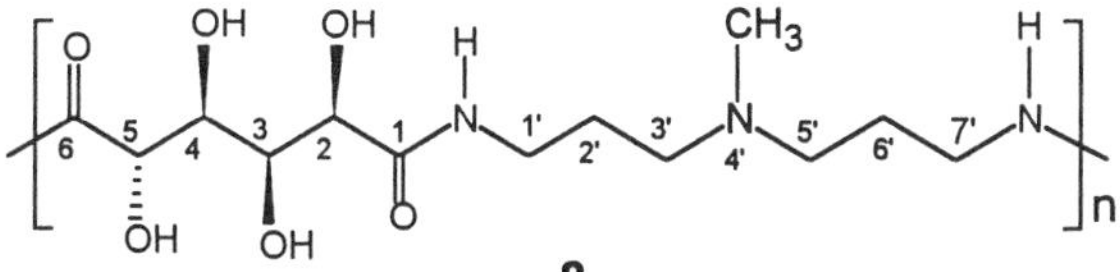

8

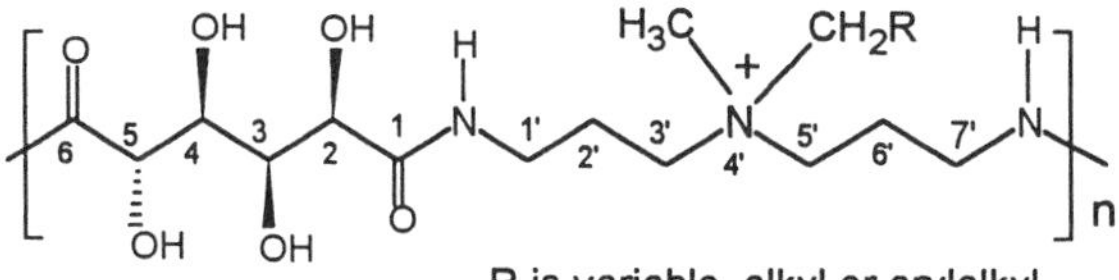

R is variable, alkyl or arylalkyl

9

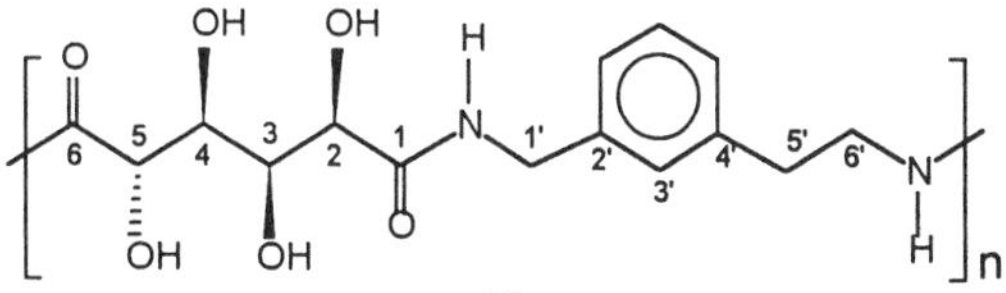

10

Figure 3. Examples of "Hydroxylated Nylons" prepared from monopotassium D-glucarate

Table. ¹H NMR Chemical Shifts and Coupling Constants of Dialkyl D-Glucaramides and Poly(hexamethylene D-glucaramide) in Trifluoroacetic Acid *

	H-2	$J_{2,3}$	H-3	$J_{3,4}$	H-4	$J_{4,5}$	H-5
DMGA	4.91	3.10	4.72	1.69	4.58	6.15	4.89
DHGA	4.94	3.27	4.74	1.76	4.58	6.46	4.89
PHGA	4.92	2.79	4.74	1.36	4.57	6.13	4.87

DMGA - Dimethyl D-glucaramide;   DHGA - Dihexyl D-glucaramide;   PHGA - Poly(hexamethylene D-glucaramide).
* Reprinted in part with permission from ref. 18, Copyright 1993.

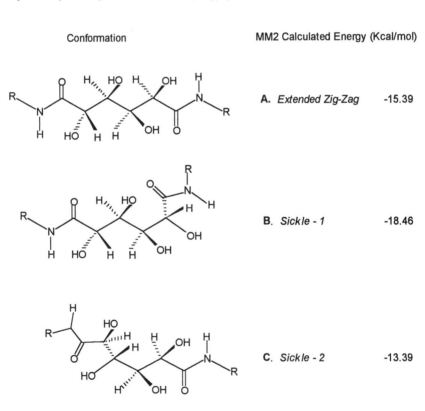

Conformation	MM2 Calculated Energy (Kcal/mol)
A. *Extended Zig-Zag*	-15.39
B. *Sickle - 1*	-18.46
C. *Sickle - 2*	-13.39

Figure 4. Conformation and MM2 Calculated Energy of Dihexyl D-Glucaramide

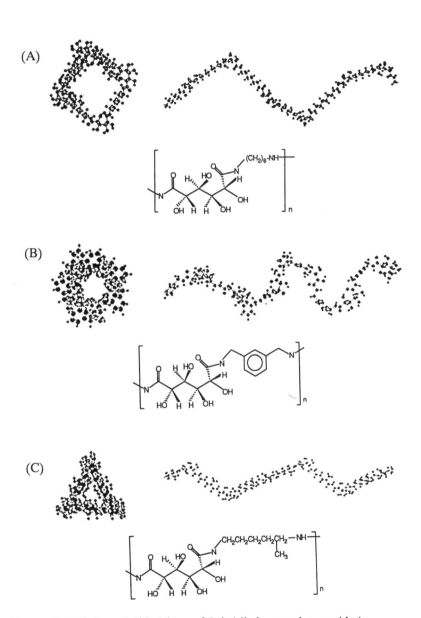

Figure 5. End On and Side Views of Poly(alkylene D-glucaramides).
A) Poly(hexamethylene D-glucaramide); B) Poly(m-xylylene D-glucaramide);
C) Poly(2-methylpentamethylene D-glucaramide).

*Reproduced in part with permission from ref 18, Copyright 1993 Division of Polymer Chemistry, ACS.

glucaramide) and **C**, poly(2-methylpentamethylene D-glucaramide).   What is clear from comparison of these three poly(D-glucaramides) is that they all form somewhat helical type structures, but with different hole shapes.   For **A**, and other poly(alkylene D-glucaramides) with an even number of methylene units in the diamine monomer unit, the hole is somewhat square.   For the the arylalkylene copolymer **B**, the hole is round and for polymer **C**, with an odd number (five) of methylene units in the diamine unit, the hole is triangular.   These computer models are of stereoregular *head,tail-head,tail* poly(D-glucaramides) (*3*),   rather than those with   randomly oriented repeating D-glucaramide monomers units.   However the actual secondary structure differences of such polymers might differ, these studies point out that secondary structure is heavily influenced by the structures of the diamine monomer units.   Consequently, such secondary structure differences are likely to be reflected by differences in physical properties of the polymers.

Literature Cited

1. Benicewicz, B. C.; Shalaby, S. W.; Clemow, A. J. T.; Oser, Z. in *Agricutural and Synthetic Polymers - Biodegradability and Utilization*; Glass, J. E. and Swift, G.; *ACS Symposium Series 433*; Washington, D.C., 1990; pp 161-173.
2. Tsai, T. L.; Sanville, C. Y.; Coleman, R. D.; Schertz, W. W. *U.S. Patent 504223 A0*, June 15, 1992.
3. Chen, L. *Ph. D. Dissertation*, The University of Alabama at Birmingham, 1992.
4. Kiely, D. E.; Chen, L.; Lin, T-H. "Hydroxylated Nylons Based on Unprotected Esterified D-Glucaric Acid by Simple Condensation Reactions", in press, *J. Am. Chem. Soc.* 1994.
5. Ogata, N.; Hosoda, Y. *J. Polym. Sci., Polym. Lett. Ed.*, **1974**, *12*, 355.
6. Ogata, N.; Sanui, K.; Iijima, K. *J. Polym. Sci., Polym. Chem. Ed.*, **1973**, *11*, 1095.
7. Ogata, N.; Okamoto, S. *J. Polym. Sci., Polym. Chem. Ed.*,**1973**, *11*, 2537.
8. Ogata, N.; Hosoda, Y. *J. Polym. Sci., Polym. Chem. Ed.*, **1975**, *13*, 1793 .
9. Ogata, N.;, *Polym. Prepr.*, **1976**, *17*, 151.
10. Ogata, N.; Sanui, K. *J. Polym. Sci., Polym. Chem. Ed.*, **1977**, *15*, 1523.
11. Ogata, N.; Sanui, K.; Nakamura, H.; Kuwahara, M. *J. Polym. Sci., Polym. Chem. Ed.*, **1980**, *18*, 939.
12. Hoagland, P. D. *Carbohydr. Res.*, **1981**, *98*, 203.
13. Hoagland, P. D.; Pessen, H.; McDonald, G. G. *J. Carbohydr. Chem.*, **1987**, *6*, 495.
14. Chen, L.; Kiely, D. E. "D-Glucaric Acid Ester/Lactones Used in Condensation Polymerization to Produce Hydroxylated Nylons - A Qualititative Equilibrium Study in Acidic and Basic Alcohol Solutions" accepted for publication in *J. Carbohydr. Chem.*, **1994**.
15. Lin, T-H. *Ph. D. Dissertation*, The University of Alabama at Birmingham, 1987.
16. Kiely, D. E.; Lin, T-H. *U.S. Patent 4,833,230*, May 23, 1989.
17. Hashimoto, K.; Okada, M.; Honjou, N. *Makromol. Chem., Rapid Comm.*, **1990**, *11*, 393.
18. Chen, L.; Kiely, D. E., *Polymer Prepr.*, **1993**, *34*, 550.
19. Chen, L.; Haraden, B; Kane, R. W.; Kiely, D. E.; Rowland, S. in *Computer Modeling of Carbohydrate Molecules*; French, A. D. and Brady, J. W., Ed.; *ACS Synposium Series 430*; American Chemical Society: Washington, D. C., 1990; pp 141-151.

RECEIVED May 24, 1994

POLYAMIDES, PROTEINS, POLYESTERS, AND RUBBERS

Chapter 11

Synthesis of Functionalized Targeted Polyamides

P. M. Mungara and K. E. Gonsalves

Polymer Science Program, Institute of Materials Science, U–136 & Department of Chemistry, University of Connecticut, Storrs, CT 06269

A synthetic approach to polyamides containing the tyrosine-leucine linkage is presented. The diphenyl phosphoryl azide, (DPPA), coupling technique was utilized to synthesize the monomers. Solution polymerization of monomer I, tyrosylleucyliminohexamethylene-iminoleucyltyrosine, with adipoyl and sebacoyl chlorides gave polymers with intrinsic viscosities of 0.18 dL/g and 0.13 dL/g respectively in 90% formic acid. Interfacial polymerization of the monomer yielded a fibrous crosslinked material which was found to be a poly(amide-ester) by IR spectroscopic analysis. In another reaction, monomer II, β–alanyltyrosylleucyl-β–alanine, was polymerized using DPPA and triethylamine to give a polyamide with an intrinsic viscosity of 0.07 dL/g in 90% formic acid. A synthetic route to monodispersed alanine peptides is also presented.

Polyamides containing the naturally occurring α-L-amino acid linkages belong to a class of potentially biodegradable polymers whose applications are numerous, especially in agriculture and the biomedical field (1). In the latter, polypeptides have had considerable success as controlled drug delivery systems, degradable sutures and artificial skin substitute(1-3).

In this paper, the synthetic approach to polyamides containing α-L-amino acids is presented. The first part deals with the synthetic method utilizing diphenyl phosphoryl azide, (DPPA), to incorporate the dipeptide, tyrosine-leucine(Tyr-Leu) in polyamides. The tyrosine-leucine bond is targeted for degradation via enzymes such as chymotrypsin, thermolysin, subtilisin and aspergillopeptidase A. In the second part, a rapid method for making monodispersed polypeptide is presented. Here, Fmoc-Ala-Cl and Benzotriazolyloxytris(dimethylamino)phosphonium hexafluorophosphate, (BOP), coupling techniques were combined to synthesize alanine peptides of upto six units, which are targeted for use in controlled drug delivery systems.

Experimental

Materials. The amino acid derivatives and the Castro's reagent, benzotriazolyloxy-tris(dimethylamino)phosphonium hexafluorophosphate, (BOP), were obtained from Advanced ChemTech Inc., Louisville, KY, and used without further purification. DPPA, obtained from Aldrich Chemical Company, was purified by distilling under reduced pressure. Adipoyl and sebacoyl chlorides were distilled under reduced pressure while hexamethylenediamine was purified by vacuum sublimation. Triethylamine was dried over CaH_2 and distilled at atmospheric pressure. Anhydrous DMF was obtained by drying over BaO and distilling under reduced pressure.

Measurements. The intrinsic viscosities of the polymers were measured in a 0.5 g/dL solution of formic acid at 25 °C. The infrared spectra of the compounds (KBr pellet, or film cast from $CHCl_3$) were recorded on a Nicolet 60SX FTIR spectrometer, while the ^1H NMR spectra were obtained on an IBM AF-270 NMR spectrometer (270 MHz), with CF_3COOD, $CDCl_3$ or DMSO-d_6 as the solvents . The molecular weights of the monomers and the alanine peptides were determined using fast atom bombardment (FAB) mass spectrometry on a Kratos MS50RF high resolution magnetic sector mass spectrometer.

Results and Discussion

Tyr-Leu Polymers. Scheme I gives the general outline of synthesizing adipoyl or sebacoyl polymers containing tyrosylleucyliminohexamethyleneiminoleucyl-tyrosine (Tyr-Leu-H-Leu-Tyr) unit. Tert-butyloxycarbonyl-L-tyrosine (BOC-Tyr-OH) was coupled with L-leucine methyl ester.HCl (H-Leu-OMe.HCl) using DPPA and triethylamine in DMF (4-7). The methyl ester protecting group was deblocked using aqueous NaOH. The incorporation of hexamethylenediamine unit was achieved by reacting 1 equivalent of the amino protected dipeptide, BOC-Tyr-Leu-OH with 0.5 equivalent of the diamine using DPPA. The resulting peptide, BOC-Tyr-Leu-NH$(CH_2)_6$NH-Leu-Tyr-BOC was reacted with anhydrous trifluoroacetic acid (TFA) at room temperature to give monomer **I**.

The ^1H NMR of the monomer in deuterated trifluoroacetic acid (CF_3COOD) showed peaks at δ (ppm); 0.81 (m, 12H, $(CH_3)_2$, Leu), 1.35-1.58 (m, 14H, CHCH$_2$, Leu, CH$_2$, diamine), 3.1-3.2 (m , 8H, CH$_2$NH, diamine, CH$_2$, Tyr), 4.5-4.6 (t, 4H , CH-N, Tyr, Leu) and 6.8-7.0 (d, 8H, aromatic H, Tyr). The IR spectrum (KBr pellet) showed a broad peak between 3500-3200 cm^{-1} (O-H stretching) peaks at 3298 cm^{-1} (N-H stretching), 3080 cm^{-1} (C-H stretching, aromatic), 2930 and 2858 cm^{-1} (C-H stretching aliphatic), 1652 cm^{-1} (amide I) and 1521 cm^{-1} (amide II).

Solution polymerization was achieved by reacting a chloroform solution of monomer **I** and triethylamine with the chloroform solutions of either adipoyl or sebacoyl chlorides to give the corresponding polymers (8). The adipoyl polymer (PTLA, 0.52 g, yield, 52.2%) had an intrinsic viscosity, [η], of 0.18 dL/g while that of sebacoyl (PTLS, 0.25 g yield, 37%) had an intrinsic viscosity of 0.13 dL/g in 90% formic acid.

The ^1H NMR of the adipoyl polymer in CF_3COOD had peaks at δ(ppm); 0.83 (d, 12H, $(CH_3)_2$, Leu), 1.3-1.5 (m, 18H, CHCH$_2$, Leu, CH$_2$, diamine, adipoyl), 2.3 (m, 4H, CH$_2$CO, adipoyl), 2.96-3.30 (m, 8H, CH$_2$, Tyr, CH$_2$N, diamine), 4.6-4.8 (m, 4H, CH-N, Tyr, Leu), and 6.8-7.0 (d, 8H, aromatic, Tyr). The IR spectrum (KBr pellet) showed peaks between 3500-3200 cm^{-1} (O-H, stretching), 3317 cm^{-1} (N-H,

Scheme I. Synthesis of Tyr-Leu-NH(CH$_2$)$_6$NH-Leu-Tyr polymers

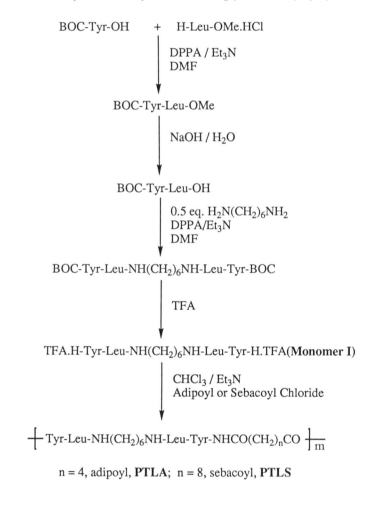

BOC-Tyr-OH   +   H-Leu-OMe.HCl

DPPA / Et$_3$N
DMF

BOC-Tyr-Leu-OMe

NaOH / H$_2$O

BOC-Tyr-Leu-OH

0.5 eq. H$_2$N(CH$_2$)$_6$NH$_2$
DPPA/Et$_3$N
DMF

BOC-Tyr-Leu-NH(CH$_2$)$_6$NH-Leu-Tyr-BOC

TFA

TFA.H-Tyr-Leu-NH(CH$_2$)$_6$NH-Leu-Tyr-H.TFA(**Monomer I**)

CHCl$_3$ / Et$_3$N
Adipoyl or Sebacoyl Chloride

$\left[\text{Tyr-Leu-NH(CH}_2)_6\text{NH-Leu-Tyr-NHCO(CH}_2)_n\text{CO} \right]_m$

n = 4, adipoyl, **PTLA**; n = 8, sebacoyl, **PTLS**

stretching), 3084 cm^{-1} (C-H stretching, aromatic), 2933 and 2853 cm^{-1} (C-H stretching, aliphatic), 1653 cm^{-1} (amide I) and 1518 cm^{-1} (amide II). Similar results were obtained for the sebacoyl polymer.

The interfacial polymerization of monomer **I** was carried out by reacting the sodium hydroxide solution of the monomer with chloroform solutions of adipoyl or sebacoyl chloride (*9*). A crosslinked poly(amide-ester) was obtained in each case. This could be attributed to the reaction of both the amino and the hydroxyl groups in the tyrosine unit with the acid chloride in alkaline medium. In order to confirm the reaction due to the hydroxyl group in the tyrosine unit, interfacial polymerization of an alkaline solution of amino protected monomer **I** and a chloroform solution of adipoyl chloride was carried out, (Scheme II). The IR spectrum of the polyester obtained was then compared with that of crosslinked poly(amide-ester). The two spectra had peaks at 1765 cm^{-1} (ester), 1656 cm^{-1} (amide I) and 1517 cm^{-1} (amide II) (*10*). The amide peaks in the polyester are due to the peptide bond originally present in the monomer. The IR spectrum of the polymer made by solution polymerization of monomer **I** in triethylamine does not show the ester peak at 1765 cm^{-1} (Figure 1). The results confirm that the hydroxyl group in tyrosine can react readily in an alkaline medium.

The synthesis of poly(β-alanyltyrosylleucyl-β-alanine), PATLA, is outlined in Scheme III. The amino acids, tert-butyloxycarbonyl-β-alanine (BOC-β-Ala-OH) and L-tyrosine methyl ester.HCl (H-Tyr-OMe.HCl) were coupled using the DPPA method as explained above. The dipeptide BOC-β-Ala-Tyr-OMe was obtained. The methyl ester protecting group was deblocked using aqueous NaOH to give the amino protected dipeptide, BOC-β-Ala-Tyr-OH. In another reaction, tert-butyloxycarbonyl-L-leucine (BOC-Leu-OH) and β-alanine methyl ester.HCl, (β-Ala-OMe.HCl) were coupled using DPPA to give the dipeptide BOC-Leu-β-Ala-OMe. The BOC protecting group was deblocked using TFA to give H-Leu-β-Ala-OMe which was utilized in segment condensation with BOC-β-Ala-Tyr-OH.

The segment condensation using DPPA and triethylamine in DMF afforded the tetrapeptide, BOC-β-Ala-Tyr-Leu-β-Ala-OMe. The NMR of the tetrapeptide in CF$_3$COOD showed peaks at δ (ppm); 0.81 (d, 6H, (CH$_3$)$_2$, Leu), 1.2-1.5 (m, 3H, CH$_2$CH, Leu), 2.7 (t, 2H , CH$_2$CO, β-Ala), 2.98 (d , 2H, CH$_2$, Tyr), 3.6 (t, 2H, CH$_2$-N, β-Ala), 4.51-4.70 (t, 2H, CH-N, Tyr, Leu), and 6.8-7.0 (d, 4H, aromatic H, Tyr).

Hydrolysis of the methyl ester protecting group of the tetrapeptide using aqueous NaOH followed by the deblocking of the BOC group with TFA afforded the TFA salt of monomer **II**, β-Ala-Tyr-Leu-β-Ala. Poly(β-Ala-Tyr-Leu-β-Ala) was obtained by dissolving monomer **II** in DMF and reacting with DPPA in the presence of triethylamine for 2 days (*11*). The polymer (PATLA, 0.50 g, yield, 55.4%) had an intrinsic viscosity, [η], of 0.07 dL/g in 90% formic acid.

The ^{1}H NMR of the polymer in CF$_3$COOD gave peaks at δ(ppm); 0.81 (6H, (CH$_3$)$_2$, Leu), 1.5 (3H, CH$_2$-CH, Leu), 2.7 (4H, CH$_2$-CO, β-Ala), 2.97 (2H, CH$_2$, Tyr), 3.56 (4H CH$_2$-N, β-Ala), 4.50-4.75 (2H, CH-N, Leu, Tyr) and 6.8-7.0 (4H, aromatic H, Tyr) . The IR spectrum (KBr pellet), showed peaks between 3500-3200 cm^{-1} (O-H stretching), 3298 cm^{-1} (N-H stretching), 3082 cm^{-1} (C-H stretching, aromatic) , 2955 and 2858 cm^{-1} (C-H stretching, aliphatic), 1670 cm^{-1} (amide I) and 1516 cm^{-1} (amide II). PATLA-**II** is the polymer made by repolymerizing

Scheme II. Synthesis of peptide polyester by interfacial method

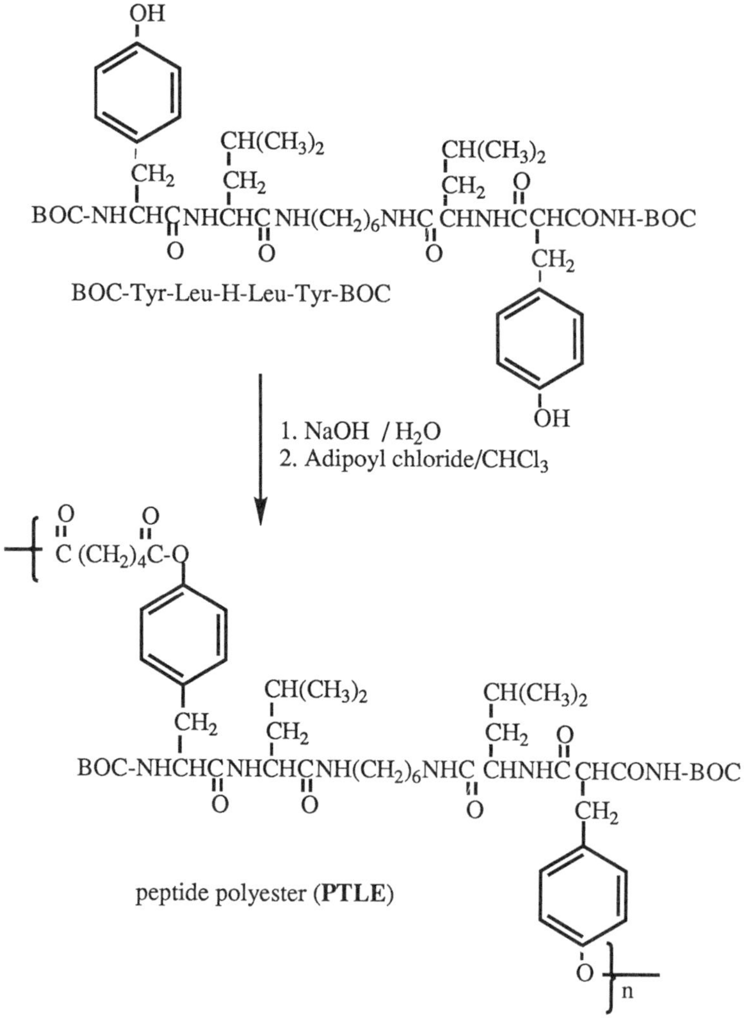

BOC-NHCHCNHCHC NH(CH$_2$)$_6$NHC CHNHCCHCONH-BOC

BOC-Tyr-Leu-H-Leu-Tyr-BOC

1. NaOH / H$_2$O
2. Adipoyl chloride/CHCl$_3$

peptide polyester (**PTLE**)

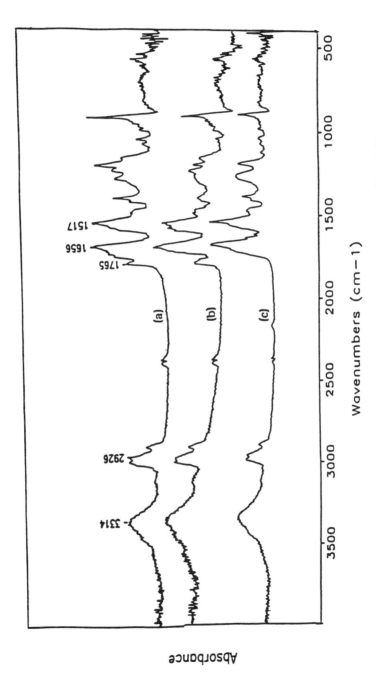

Figure 1. Infrared spectra of polymers: (a) poly(amide-ester), cross-linked, (b) peptide polyester, PTLE, (c) polyamide, PTLA.

Scheme III. Synthesis of β-Ala-Tyr-Leu-β-Ala polymer

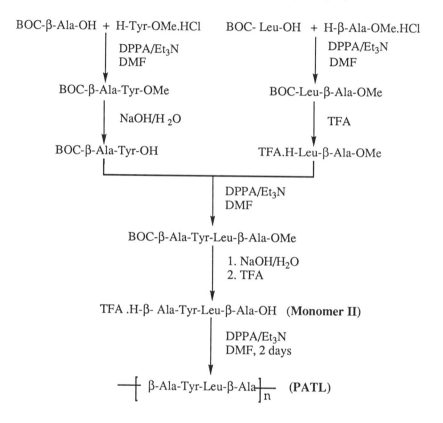

Table I. Yields, T_m and GPC Data of the Tyr-Leu Polymers /Oligomers

	Sample				
	PTLA	PTLS	PTLE	PATLA	PATLA-II
Yield, %	52.2	37	44.4	55.4	55.3
M_w	14200	20800	3500	5300	6500
M_n	7200	8100	2500	2900	5700
M_w/M_n	1.97	2.57	1.40	1.83	2.89
T_m(°C)	230	160	131	a	138

a not determined

PATLA under similar condition as before. As shown in Table I, there was some improvement in the molecular weight although the polydispersity also increased.

The result from Table I indicate medium to low yields of the polymers. This could be attributed mainly to the insolubility of the polymers in the solvent medium used in the case of solution polymerization or insufficient stirring speed in the interfacial method. Besides, the bulky nature of the monomers cannot be discounted as contributing to steric hindrance to the reactions resulting into low yields and low molecular weights.

Alanine Peptides. The synthesis scheme utilizing Fmoc-Ala-Cl is given in Scheme IV and is essentially that of Carpino et al. (*12*). Fmoc-Ala-Cl was synthesized by refluxing Fmoc-Ala-OH (6.0 g) with $SOCl_2$ (12 mL) using CH_2Cl_2 (30mL) as the solvent. The amino acid chloride was purified by washing several times in CH_2Cl_2 and evaporating off the solvent under reduced pressure to ensure complete removal of the $SOCl_2$. Recrystallization of the product from hexane-CH_2Cl_2 gave Fmoc-Ala-Cl (5.90 g, yield, 92.8% , m.p. 112- 114 ^{o} C) which was used in acylation without further purification. A methylene chloride solution of Fmoc-Ala-Cl was interfacially coupled with H-Ala-OtBu.HCl in 5% Na_2CO_3. The reaction mixture was easily purified to afford the dipeptide, Fmoc-Ala-Ala-OtBu (4.0 g, 91.3 % m.p. 161-162 ^{o}C).

The ^{1}H NMR of the compound in CDCl$_3$ showed peaks at δ (ppm); 1.2 (d, 6H, CH-CH$_3$), 1.4 (s, 9H, C(CH$_3$)$_3$), 4.2-4.4 (m, 5H, CH-CH$_3$ and CH-CH$_2$O), 5.3 (d, 1H, NH), 6.3 (d, 1H, NH), and 7.3-7.8 (m, 8H, aryl). IR spectrum (film, solvent; CHCl3) gave peaks at 3286 cm^{-1} (N-H stretching), 3065 cm-1 (aryl C-H stretching), 2963 cm^{-1} (aliphatic C-H stretching), 1739 cm^{-1} (C=O, CH$_2$-O-C=O), and 1670 cm^{-1} (C=O, amide).

The Fmoc protecting group of the dipeptide was deblocked using Et$_2$NH in acetonitrile and the resulting carboxyl protected dipeptide was acylated with Fmoc-Ala-Cl to give the tripeptide, Fmoc-Ala-Ala-Ala-OtBu (2.0 g yield 78.4% m.p. 189-190 ^{o}C) .

NMR, δ (pmm), CDCl3; 1.2 (d, 9H, CH-CH$_3$), 1.5 (s, 9H, C(CH$_3$)$_3$), 4.2-4.4 (m, 6H, N-CH-CO and CH-CH$_2$-O), 5.41 (broad s, 1H, NH), 6.6 (broad s, 2H, N-H), 7.3-7.8 (m, aryl). IR (film, solvent; CDCl3, cm^{-1}) ; 3286 (N-H stretching), 3050 (aryl C-H stretching) 2946 (aliphatic C-H stretching), 1722 (CO stretching, CH$_2$-O-C=O), 1 520 (C=O stretching, amide).

The peptides obtained by the rapid Fmoc-Ala-Cl acylation technique were utilized in a segment condensation reaction using BOP coupling reagent and diisopropylethylamine, DIEA (*13,14*) . Scheme V gives the outline of the synthetic process leading to the tetrapeptide, Fmoc-Ala-Ala-Ala-Ala-OtBu. Here the Fmoc group was deblocked using Et$_2$NH to give the dipeptide, H-Ala-Ala-OtBu. On the other hand, TFA was used to deblock the tertiary butyl group to give Fmoc-Ala-Ala-OH. The two dipeptides were then coupled at room temperature using BOP and DIEA in DMF to afford the tetrapeptide, Fmoc-Ala-Ala-Ala-Ala-OtBu (1.3 g, yield, 89.6%, decompose at 190 ^{o}C).

The ^{1}H NMR of this peptide showed peaks at δ (ppm), (DMSO-d$_6$); 1.2 (d, 12H, CH-CH$_3$), 1.4 (s, 9H, C(CH$_3$)$_3$), 4.1-4.3 (m, 7H, N-CH-C=O and CH-CH$_2$-O), 7.3-8.3 (m, 12H, N-H and aryl). IR (KBr pellet, cm^{-1}); 3295 (N-H stretching), 3050 (aryl C-H stretching), 2963 and 2912 (aliphatic C-H stretching), 1722 (s, C=O stretching, CH$_2$-O-C=O), 1620 (C=O, amide). Similar coupling technique as

Scheme IV. Fmoc-Ala-Cl acylation technique

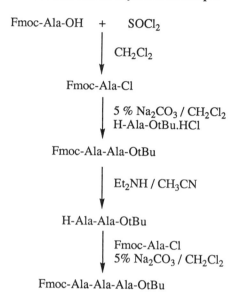

Fmoc-Ala-OH   +   SOCl$_2$

| CH$_2$Cl$_2$

Fmoc-Ala-Cl

| 5 % Na$_2$CO$_3$ / CH$_2$Cl$_2$
| H-Ala-OtBu.HCl

Fmoc-Ala-Ala-OtBu

| Et$_2$NH / CH$_3$CN

H-Ala-Ala-OtBu

| Fmoc-Ala-Cl
| 5% Na$_2$CO$_3$ / CH$_2$Cl$_2$

Fmoc-Ala-Ala-Ala-OtBu

Scheme V. BOP 2 + 2 segment coupling *

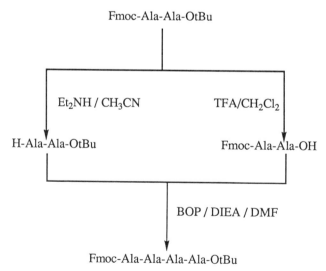

Fmoc-Ala-Ala-OtBu

Et$_2$NH / CH$_3$CN   TFA/CH$_2$Cl$_2$

H-Ala-Ala-OtBu   Fmoc-Ala-Ala-OH

BOP / DIEA / DMF

Fmoc-Ala-Ala-Ala-Ala-OtBu

*The same procedure was followed to make hexa-alanine

described for the tetrapeptide was used to synthesize Fmoc-Ala$_6$-OtBu, (0.87 g yield, 84.3 %, m.p. 236 °C dec.), from trialanine.

NMR, δ (ppm), DMSO-d$_6$; 1.2 (d, 18H, CH-CH$_3$), 1.4 (s, 9H, C(CH$_3$)$_3$), 4.1-4.3 (m, 9H, N-CH-CO and CH-CH$_2$-O), 7.3-8.3 (m, 14H, N-H and aryl). IR (KBr pellet, cm^{-1}); 3330 (N-H stretching), 3050 (aryl C-H stretching) 2967 and 2933 (aliphatic C-H stretching), 1733 (C=O stretching, CH$_2$-O-C=O), 1633 (s, C=O amide).

Table II gives a summary of the yields and the melting points of the alanine peptides synthesized by both Fmoc-Ala-Cl acylation and BOP coupling technique.

Table II.   Yields and M.p. of the Alanine Peptides

Sample	Yield(%)	m.p. (°C)
Fmoc-dialanine-OtBu	91.3	161-162
Fmoc-trialanine-OtBu	78.4	189-190
Fmoc-tetra-alanine-OtBu	89.6	190(dec.)
Fmoc-hexa-alanine-OtBu	84.3	236 (dec.)

Conclusions

These studies indicate that the DPPA technique has been successful in synthesizing peptides containing the tyrosine-leucine linkage, a potential target for enzymatic degradation. These peptides can be utilized in both solution and interfacial, polymerization (Schemes I & II). Direct polymerization is also possible using DPPA (Scheme III). From Scheme II, it has been demonstrated that a functionalized polymer can be synthesized e.g., polyamides containing pendant hydroxyl groups or conversely polyesters containing pendant amino groups. Modifications to obtain higher molecular weight polymers are underway.

The Fmoc-Ala-Cl and BOP coupling techniques are convenient for synthesizing the alanine peptides at room temperature using very mild conditions. The yields obtained are also high. Due to the encouranging results, we intend to synthesize longer alanine peptides which can serve as model compounds in the study of peptide structures.

Literature Cited

1. Kumar, G. S. *Biodegradable Polymers, Prospect and Progress;* Marcel Dekker, Inc. : New York, 1987.
2. Anderson, J. M.; Gibbons, D. F.; Martin, R. L.; Hiltner, A.; Woods, R. J. *Biomed. Mater. Res. Symp.* **1974**, *5*(1), 197.
3. Hayashi, T.; Iwatsuki, M. *Biopolymers* **1990**, *29*, 549.
4. Gonsalves, K. E.; Mungara, P. M. *Chem. Mater.* **1993**, *5*, 1242.
5. Shioiri, T.; Ninomiya, K.; Yamada, S. *J. Amer. Chem. Soc.* **1972**, *94*, 6203.
6. Yamada, S.; Ikota, N.; Shioiri, T. *J. Amer. Chem. Soc.* **1975**, *97*, 7175.
7. Gonsalves, K. E.; Mungara, P. M. *Polym. Commun.* **1994**, *35*(3), 663.
8. Morgan, P. W.; Kwolek, S. L. *J. Polymer Sci.* **1964**, *A2*, 185 .
9. Gonsalves, K. E.; Chen, X.; Wong, T. K. *J. Mater. Chem.* **1991**, *1*(4), 643.

10. Gonsalves, K. E.; Chen, X.; Cameron, J. A. *Macromolecules* **1992**, *25,* 3309.
11. Nishi, N.; Tsunemi, M.; Hayasaka, H.; Nakajima, B.; Tokura, S. *Makromol. Chem.* **1991**, *192* , 1789.
12. Carpino, L.A.; Aalaee, D. S.; Beyermann, M.; Bienert, M.; Niedrich, H. *J. Org. Chem.* **1990**, *55*, 721.
13. Ten Kortenaar, P. B. W.; Kruse, J.; Hemminga, M. A.; Tesser, G. I. *Int. J. Peptide Protein Res.* **1986**, *27*, 401.
14. Fehrentz, J. A.; Seyer, R.; Heitz, A.; Fulcrand, P.; Castro B. ; Corvol, P. *Int. J. Peptide Protein Res.* **1986**, *28*, 620.

RECEIVED May 24, 1994

Chapter 12

Extraction and Characterization of "Chrome-Free" Protein from Chromium-Containing Collagenous Waste Generated in the Leather Industry

M. M. Taylor, E. J. Diefendorf, C. J. Thompson, E. M. Brown, and W. N. Marmer

Eastern Regional Research Center, Agricultural Research Service, U.S. Department of Agriculture, 600 East Mermaid Lane, Philadelphia, PA 19118

In the United States, almost 56,000 metric tons of chromium-containing solid waste are generated by the leather industry each year, and approximately ten times this amount is generated world-wide. Environmental concerns and escalating landfill costs are becoming increasingly serious problems to the leather industry and alternative disposal methods are needed. We have developed a process in which this waste is first treated with alkali to extract a high molecular weight gelable collagen protein. The sludge that remains is further treated with enzymes to recover a lower molecular weight protein hydrolyzate and a recyclable chromium product. The recovered protein fractions, practically devoid of chromium, could be used in a wide range of products, including adhesives, cosmetics, films, animal feed and fertilizer. The isolated chromium can be chemically treated and recycled into the tanning process.

Historically, shavings, trimmings and splits from the chrome tanning of hides and skins have been disposed of in landfills. Recently, tighter local restrictions have caused the tanning industry to seek alternatives to dumping. During the past 25 years, many investigators have developed some rather innovative methods to treat this waste product. Alkali hydrolysis has been one of the most investigated processes. For example, researchers used calcium hydroxide and steam for the purpose of chrome recovery and isolation of a protein fraction (1-2). Studies have been done with sodium hydroxide and pressure to improve the efficiency of the reaction (3), with ammonia to obtain fertilizers and with sodium carbonate and sodium hydroxide combinations to produce coagulants for natural rubber and leveling agents for leather dyeing (4).

Several investigators have hydrolyzed the waste products from chrome tanning with sulfuric acid and used the chromium-containing hydrolysate as a retanning agent (5-6) or, after precipitation of the chromium, used the isolated amino acids as an animal feed supplement (7). The hydrolysates also may be used to produce fatliquors, surfactants and fillers for leather manufacture (8). Others have hydrolyzed with organic acids to obtain oligopeptides (9); acrylic acids have been used and the resulting hydrolysate was copolymerized with vinyl monomers to give fillers for leathers (10).

Many investigators have recovered the chromium by wet air oxidation (11), peroxide treatments (12), and incineration at a variety of temperatures (13-15). Chromium with the +6 oxidation state would be generated in these reactions and a reduction step would be needed.

Reaction of the chromium-containing waste product with monomers and polymers has been carried out by a number of investigators. The leather scraps have been reacted with polyisocyanates to make insulators and building materials (16). The substrate has been grafted with hydrophilic acrylates to make fibrous sheets (17), polymerized to make molded products fillers for leather (18), and, with vinyl acetate, formed into sheets for shoe soles (19).

Other uses include addition of the recovered chrome to cements and mortars (20). The waste product has been used in the manufacture of composite sheets for footwear (21). Leather substitutes have been made by a papermaking method (22) and mixtures of chrome leather fibers and cellulose pulp have been used as a paper substitute (23). Several researchers have detanned the chrome product for the purpose of gelatin preparation (24-25) and others have been able to isolate collagen fibers (26-27).

We have developed a process that can help the leather industry in solving this potentially difficult waste disposal problem. In this process, the chrome waste is treated with alkaline proteolytic enzymes at moderate temperatures for a short period of time. The process is unique because the pH at which the reaction takes place (8.3 to 10.5) prevents the chromium from going into solution, thus averting the poisoning of the enzyme by chromium and enabling the recovery of chromium as $Cr(OH)_3$ by filtration. The resulting protein solution may have commercial use as a feed or a fertilizer or could be discarded as sanitary sewage. The isolated residue containing chromium and organic matter (chrome cake) has the potential to be recycled into the tanning process by treatment with sulfuric acid. The tanning industry has begun collaborative efforts with us to assess the process on a commercial scale.

It had been documented (28-39) that chromium-containing waste can be treated enzymically, but only after denaturation of the collagen. The methods developed at this laboratory demonstrated that the collagen may be denatured in the presence of alkali at moderate temperatures and thus the direct addition of the enzyme to shavings already subjected to moderate pretreatment temperatures may be made. Maintenance of these temperatures throughout the entire digestion process eliminates the need to cool the reaction mixture.

In preliminary investigations using calcium hydroxide to control the pH (37-39), 78% solubilization of the shavings was achieved when 6% (based on wet weight of shavings) of an alkaline proteolytic enzyme was used for hydrolysis. When magnesium oxide was used in conjunction with other alkaline agents *(40-43)*, higher solubilization of protein was achieved with lower amounts of enzyme than previously reported, thus making the treatment more cost-effective.

More recently, we have found that if a two step process is used, a *gelable* protein product can be obtained that should provide a higher economic return. In this process, which is covered by a new patent *(44)*, the chromium waste is treated with alkaline agents for six hours at 70-72°C and then filtered to recover a gelable protein. The chromium-containing sludge that remains is then treated with the bacterial enzyme as in the original process, resulting in a protein hydrolysate fraction and a chrome cake that can be chemically treated and subsequently recycled.

The protein products that result from these two treatments have many possible uses. Because of its high nitrogen content, the isolated protein has potential as a fertilizer and as an animal feed additive. The gelable protein has potential use in cosmetics, adhesives, printing or photography.

Materials

Alcalase (alkaline protease) was obtained from Novo Nordisk Bioindustrials, Inc. (Danbury, CT). It is a proteolytic enzyme with optimal activity at pH 8.3-9.0 and 55-65°C. It is supplied both as a granular solid (adsorbed onto an inert carrier and standardized to contain 2.0 Anson Units/g (AU/g)), and as a liquid (standardized to contain 2.5 AU/g). Liquid Alcalase was used in these experiments.

Pluronic 25R2, a non-ionic surfactant, was obtained from BASF (Parsippany, NJ). Magnesium oxide was obtained from J.T. Baker Chemical Co. (Phillipsburg, NJ) and from Martin Marietta Magnesia Specialties (Hunt Valley, MD) as MagChem 50. Sodium hydroxide (50% solution), potassium hydroxide, sodium carbonate and potassium carbonate were obtained from J.T. Baker Chemical Co. (Phillipsburg, NJ).

Procedure

Recovery of Hydrolyzed Protein Products. Chromium-containing leather waste was obtained from commercial tanneries. Sample A shavings came from a conventional chrome tannage. Sample B shavings came from a tannage in which a high exhaust chrome treatment had been used in order to reduce the chromium in the effluent. Sample C shavings came from a tannage in which the final pH was slightly more acidic (pH 3.6) than other chrome offal investigated (pH 3.8-4.2).

Each (11.5 kg) of the shavings samples (A, B and C) was pretreated with agitation at 67-69°C in 56 L (500% float) of water for two hours. Bench type experiments determined the best pretreatment for each individual sample prior to the pilot scale runs. This pretreatment step is necessary to obtain the pH that will be optimal for the enzymic digestion. Thus, Sample A was pretreated with 575 g magnesium oxide, and Sample B with 345 g NaOH and 230 g magnesium oxide.

After several preliminary bench experiments, it was found that Sample C needed to be treated with 690 g magnesium oxide (C-1). Because of its acidity, another portion of this sample was pretreated with 345 g NaOH and 345 g magnesium oxide (C-2). The enzyme (345 g) was added in three feeds (172.5 g in each feed) to each of the four reactions, over a three hour period. Upon completion of the digestion (67-69°C for 3 hr), the sample was pumped from the reaction vessel and allowed to settle overnight. The protein hydrolysate layer was decanted and the settled chromium layer was filtered through Whatman #1 filter paper. An aliquot of each protein layer was stored at 4°C. The unwashed chrome cake was collected and it, too, was stored at 4°C.

Recovery of Gelable and Hydrolyzed Protein Products. Chromium-containing leather waste was obtained from a commercial tannery. Two samples (A and B) were received over a four month time period.

Two hundred grams of either of the chrome shavings samples (A and B) were shaken in 1 L of water (500% float), 0.2 g of a non-ionic surfactant and the appropriate alkali at 70-72°C for six hours. The samples were filtered hot through Whatman #1 filter paper. The chrome sludge and the filtered gelable protein solutions were stored at 4°C. The chrome sludge was warmed to room temperature and 200 mL water (100% float) and 0.2 g non-ionic surfactant were added. The samples were shaken at 70-72°C for 1.5 hrs. The pH was adjusted with magnesium oxide to optimal pH for the enzyme. The enzyme (0.2 g) was added and the samples were shaken at 70-72°C for 3.5 hrs. The solutions were filtered hot through Whatman #1 filter paper and the hydrolyzed protein solutions were stored at 4°C. The chrome cake was air dried.

Treatment of Chrome Cake. One gram of air dried chrome cake was dissolved in 50 mL of 3.6N (10%) sulfuric acid. The pH was < 1.0. The pH of the solution was slowly raised to 1.85 - 2.00 with 0.25N NaOH and a flocculent precipitate formed. The solution was heated for several minutes at 60°C, was allowed to stand overnight and was then filtered. The residue was washed with 0.01N sulfuric acid to remove trapped chromium. The residue was dried overnight at 60°C and then weighed; the percent residue was calculated. The residue was ashed at 600°C in a muffle furnace and percent ash and volatile solids were calculated.

Analyses

The chrome shavings were analyzed for moisture, ash, total solids, total ash, Total Kjeldahl Nitrogen (TKN), fat, calcium, magnesium, and chromium as described in a previous publication (45). Amino acid analyses were carried out on a Beckman Model 119CL Analyzer.

Protein molecular weights were estimated by SDS-PAGE (polyacrylamide gel electrophoresis in sodium dodecylsulfate) (46) using a PhastSystem by Pharmacia.

The instrument used to measure the Bloom value was the TA.XT2 Texture Analyzer from Texture Technologies Corporation, Scarsdale, NY. Gel strengths were measured by Bloom determinations *(47)*. The dried gelatin (7.5 g) was weighed into a Bloom jar and 105 mL of water was added, to give a 6.67% weight/weight concentration. Water was absorbed for a set period of time (10 min to overnight) and then heated in a 65°C bath for 15 minutes, cooled at room temperature for 15 min and then placed in a 10°C bath for 17-18 hr. The sample was placed under a 0.5 inch analyzer probe and the probe was driven into the sample to a depth of 4 mm at a rate of 1 mm per sec. The force required for this was expressed as the Bloom value.

Recovery of Protein Solely as the Hydrolyzed Product (the "Original" Process)

Not all chromium-containing leather waste is the same. Tanneries use different processes to tan leather. These differences are introduced not only to affect the properties of the tanned leather, but also, in some cases, to allow high chrome exhaustion of the tanning liquor for environmental reasons. The protocol for the pretreatment of these shavings must be adjusted to achieve optimal solubility. The commercial value of this process depends not only on the savings from decreased landfill fees, but also on the value of the recovered reaction products. Thus, it is important to know the chemical composition of the isolated chrome cakes.

Chrome shavings from various tanneries were analyzed for moisture, ash, chromium, nitrogen, fat, calcium, and magnesium. The results of these analyses allow a prediction of the chemical composition of the chrome cakes. Table I shows the results of these analyses. Each of the shavings contained about the same

Table I. Analyses of Chrome Shavings

Parameter % [a]	A	B	C
Moisture	53.51 ± 0.28	53.47 ± 1.04	51.47 ± 0.36
Ash [b]	14.32 ± 0.10	8.40 ± 0.48	14.95 ± 0.37
Chromic Oxide [b]	4.21 ± 0.03	4.28 ± 0.09	3.99 ± 0.11
TKN [b,c,d]	14.54 ± 0.48	14.56 ± 0.24	14.13 ± 0.16
Fat [b]	0.09 ± 0.01	1.51 ± 0.36	1.79 ± 0.22
Calcium [b]	0.34 ± 0.01	0.40 ± 0.01	0.48 ± 0.01
Magnesium [b]	0.33 ± 0.02	0.08 ± 0.01	0.16 ± 0.01

[a] N = 3 where N = number of replicates for each sample.
[b] Moisture free basis.
[c] Total Kjeldahl nitrogen.
[d] Protein content can be estimated by multiplying TKN by 5.51.

amount of moisture, from 51.5 to 53.5%. Ash content ranged from 8 to 15%. Chromic oxide content ranged from 3.99 to 4.28%. The nitrogen content was 14.1 to 14.6%, and may be correlated to the protein content of the shavings (roughly 80% on a moisture-free basis). The fat content varied from 0.1 to 1.8%. Calcium values ranged from 0.34 to 0.48% and magnesium from 0.08 to 0.33%.

The chemical composition of the recovered chrome cakes from each treatment is shown in Table II. The fat contents reflect the amount of fat found in the untreated shavings (Table I). The fat content in Sample B may also reflect the compounds that had been used in the high exhaust chrome treatment. These compounds appeared to be lipophilic, for the extracts from these samples, dark brown and viscous, were different from the other two. The cakes were not washed during filtration; the nitrogen content reflects the protein that remains and is dependent on the efficiency of the filtration process. The chromic oxide content reflects the amount of chrome in the original shavings. The magnesium content reflects the amount of magnesium used in the pretreatments. The value for calcium found in the cakes may reflect the approximately 1% calcium impurity in the magnesium oxide.

Table II. Analyses of Chrome Cakes from Enzymic Treatment
of Chrome Shavings

Parameter % [a]	A	B	C-1	C-2
Moisture	85.42 ± 0.17	85.54 ± 0.22	82.93 ± 0.60	82.53 ± 0.94
Ash [b]	35.45 ± 0.08	32.55 ± 0.49	34.14 ± 0.83	36.99 ± 0.38
Chromic Oxide [b]	7.76 ± 0.30	11.82 ± 0.54	8.74 ± 0.10	11.44 ± 0.03
TKN[b,c,d]	7.51 ± 0.09	8.40 ± 0.66	6.66 ± 0.24	8.09 ± 0.55
Fat [b]	1.37 ± 0.10	6.31 ± 0.38	4.26 ± 0.07	4.93 ± 0.06
Calcium[b]	0.35 ± 0.01	0.82 ± 0.02	0.75 ± 0.06	1.18 ± 0.08
Magnesium[b]	9.96 ± 0.12	5.00 ± 0.06	9.47 ± 0.16	6.73 ± 0.22

[a] N = 3 where N = number of replicates for each sample.
[b] Moisture free basis.
[c] Total Kjeldahl nitrogen.
[d] Protein content can be estimated by multiplying TKN by 5.51.
Reproduced with permission from reference 43. Copyright 1992 Journal of the American Leather Chemists Association.

The isolated protein hydrolysates were analyzed for chromium, total Kjeldahl nitrogen (TKN), total solids and ash (Table III). Average values for samples A, B, C-1 and C-2 show that the chromium content was less than 1 ppm. This chromium concentration is similar to the concentrations found, not only in testing of protein from pilot studies, but also the protein solution recovered from industrial scale trials. The TKN, total solids and total ash averaged about 11,000 ppm, 72,000 ppm and 8,000 ppm, respectively. The molecular weight distribution of the hydrolyzed protein ranged from 1000-3000.

Table III. Analyses of Protein Hydrolysates

Solubilization [a] with 1% Enzyme		80
Protein Hydrolysate (Liquid) [b]		
Chromium	(AV)	< 1
TKN	(AV)	11,000
Total Solids	(AV)	72,000
Total Ash	(AV)	8,000
Protein Hydrolysate (Dried) [a]		
TKN		13.8-15.0
Ash		9.7-18.9
Molecular Weight Distribution		1,000-3,000

[a] Expressed as percent.
[b] Expressed in PPM.

A composite of the amino acid analyses for each of the dried protein samples is shown in Table IV. The values are expressed as mole percent. When the profile of the protein hydrolysate is compared to the profile of collagen, the results are quite similar, suggesting that no modification of amino acids occurred during processing.

It was demonstrated that full splits and trimmings could be enzymically hydrolyzed. In this treatment, the alkali pretreatment time was extended to three hours and the temperature was increased to 70-72°C. The structure of the hides was so totally disrupted that upon addition of the enzyme, the samples were readily digested.   It was decided to apply this extended holding time and higher temperature to chrome shavings and it was found that 0.3% of an alkaline protease was successful in digesting the shavings and giving a clean cake. Thus, the amount of enzyme that was suggested previously had been reduced almost five times.

Recycling of the protein solution containing the enzyme was attempted. The enzyme did not denature after being subjected to the high temperatures and pH's and it was found that one could successfully recycle the protein solution and enzyme, not once, but four times. The salt concentration eventually became quite

Table IV. Amino Acid Composition of Hydrolyzed Protein [a]

Residue	Collagen (Type I)	Hydrolysate	Std Dev [b]
Gly	32.7	33.0	1.7
Hyp	8.6	10.0	1.2
Pro	13.0	12.5	0.5
Ala	11.4	8.4	0.6
Arg	5.2	4.8	0.3
Asp	4.6	5.1	0.1
Cys	0.0	0.0	0.0
Glu	7.5	7.7	0.3
His	0.5	0.9	0.6
Ile	1.2	1.4	0.2
Leu	2.5	2.6	0.1
Lys	2.8	2.7	0.2
Met	0.6	0.2	0.3
Phe	1.3	1.3	0.0
Ser	3.1	4.1	0.9
Thr	1.6	2.1	0.7
Tyr	0.4	0.5	0.1
Val	2.3	2.4	0.1
Total	99.3	100.0	

[a] Expressed as mole percent.
[b] Hydrolysate samples.
Reproduced with permission from reference 43. Copyright 1992 Journal of the American Leather Chemists Association.

high and the enzyme lost its activity. A 1% concentration of the enzyme initially is recommended if one is recycling.

Recovery of Gelable and Hydrolyzed Protein Products (the "New" Process).

The next important step in the investigations would be to obtain a higher molecular weight protein than was previously isolated. The original one-step process gave not only a recyclable chromium product but also gave a low return protein hydrolysate that could possibly be used as animal feed and fertilizer. The economic return from these products would not make this process viable unless the landfill fees were exorbitant or there were no outlets for disposal of this waste product. Even though it was demonstrated that this protein solution and enzyme could be recycled in order to reduce the cost of the process, a higher return from a better quality by-product would be desirable.

Extraction of gelatin from chromium leather waste has been described in the literature *(24-25)*. However, a considerable amount of chromium sludge remains after this extraction and disposal of this sludge is necessary *(48)*. A new two-step process was proposed that would isolate a gelable protein in the first step and a lower molecular weight, hydrolyzed protein after enzymic treatment of the remaining chrome sludge. A filterable and recyclable chromium product would also be obtained.

Figure 1 is a flow diagram of the new process and illustrates one of the many alkali-inducing combinations that can be used to extract the gelable protein. After isolation of the gelable protein product by filtration, the chrome sludge is prepared for enzymic hydrolysis. The pH is measured and adjusted if necessary for optimal enzyme activity and the alkaline protease is added. The reaction was carried out for 3.5 hours. The protein hydrolysate solution can be recycled to reduce enzyme costs, and a 0.3% initial feed of enzyme is recommended. If one has whole splits or large trimmings, chipping or grinding is recommended before the first step is carried out. These substrates have been dissolved in their intact state, with 1% enzyme or less, but the protein product is low molecular weight.

In the original process, the chromium-containing waste was pretreated with a variety of alkalis at 67-69°C, not only to aid in the denaturing of the collagen but also to prepare the system for the optimal pH for the enzyme. At that time magnesium oxide, calcium hydroxide and various combinations of magnesium oxide, sodium hydroxide, sodium carbonate and calcium hydroxide were used. These various agents were used so that the process could be worked into whatever chrome recycling system the tannery would be using, since all these chemicals have been used in chrome precipitation *(49-50)*.

The effect that magnesium oxide, alone and in combination with varying amounts of sodium hydroxide, sodium carbonate, potassium hydroxide and potassium carbonate, has on the chemical and physical properties of the gelable and hydrolyzed protein products has been investigated. As has been shown in previous publications *(40-41)*, careful control of the concentrations of the alkali-inducing agents will give the optimal pH range for enzyme hydrolysis and—as will be shown—the optimal range for gelable protein extraction. Also, the pH of the reaction should not fall below 8.5, for then there would be the risk of solubilizing the chromium. Shavings from different tannery processes have different pH values, ranging from 3.50 to 4.20. The shavings being used in these experiments had a pH range of 3.95 to 4.00. The concentrations of alkalis to be added were arrived at experimentally in small bench trials prior to larger scale runs.

Table V summarizes the chemical and physical properties of the gelable protein that has been extracted using various combinations of the above-mentioned alkali treatments. The percent total solids in the solutions can range from 1.75 to 4%, depending on the choice of alkali. The chromium content of the protein products can range from 0.005 to 0.0126%. These gelable protein solutions were freeze-dried to a white solid with a moisture content ranging from 4 to 13 % and

Example of U.S. Patent 5,271,912 (CIP of 5,094,946)

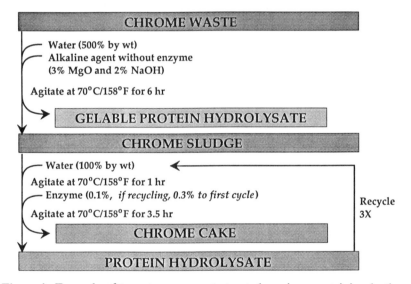

Figure 1. Example of two step process to treat chromium-containing leather waste. A variety of alkalinity-inducing agents can be used to extract gelable protein products.

Reproduced with permission from reference 56. Copyright 1992 Journal of the American Leather Chemists Association.

the ash content from 8.9 to 21%. Molecular weight distribution can range from 75,000 to more than 200,000, depending on the alkali treatment that is used. The Bloom values, or gel strengths, range from 80 to 150 g.

Table V. Characterization of Gelable Protein Products

Parameters	Range
Total Solids	1.75-4.00%
Moisture	4.00-13.00%
Ash[a]	8.90-20.00%
Chromium[a]	.005-.013%
Molecular Weight Distribution	75,000- > 200,000
Bloom Value	80-150 g

[a] Moisture-free basis (MFB).
Reproduced with permission from reference 56. Copyright 1992
Journal of the American Leather Chemists Association.

Table VI summarizes the chemical properties of the hydrolyzed protein products isolated from the chrome sludge, the character of which will vary depending on the choice of alkali. The molecular weight ranged from 10,000 to 20,000, values much higher than those protein products isolated in the original process; this reflects the small amount of enzyme used in the sludge digestion.

Table VI. Characterization of Hydrolyzed Protein Products

Parameters	Range
Total Solids	6.00-9.50%
Ash[a]	3.30-7.70%
Chromium[a]	0.0005-0.005%
Molecular Weight Distribution	10,000-20,000

[a] Moisture-free basis (MFB).
Reproduced with permission from reference 56. Copyright 1992
Journal of the American Leather Chemists Association.

Also shown are the range of total solids and total ash content of the protein solutions along with the range of the chromium concentration. The total ash content of the hydrolyzed protein products is much lower than that found in the original process.

Possible uses of the gelable protein fractions include graft polymerized products, waste-water treatment, encapsulation, powdered filler for skid resistant tires and thermal printing materials. Possible uses for the hydrolysates include fertilizer, animal feed, and adhesives.

Because there is concern that the character and quantity of the ash in the protein products would have an adverse effect on marketing of these products, the use of different proportions of alkaline agents on the ash content of the protein products was investigated. Increasing proportions of magnesium oxide resulted in lower ash content.

In Table VII, the effects of potassium hydroxide and potassium carbonate with magnesium oxide and their effect on the ash content of the gels and hydrolysates are shown. Potassium is used in fertilizers and would be advantageous if the hydrolysate product could be used in this market. Increasing the hydroxide and carbonate concentrations increases the ash content of the gelable fraction significantly. The total solids and total ash content of the hydrolysates increases with higher concentrations of hydroxide and carbonate.

Table VII. Effect of MgO and KOH or K_2CO_3 Concentrations on Character of Gelable Protein and Hydrolyzed Protein Products[a]

Parameter [b]	6% MgO	4% 1% MgO-KOH	3% 2% MgO-KOH	5% 1% Mg-CO$_3$	4% 2% Mg-CO$_3$
Gelable Protein					
Final pH	8.90	8.72	8.78	8.40	8.40
% Total Solids	2.20±0.01	2.48±0.00	2.29±0.00	2.30±0.01	2.17±0.01
% Total Ash	0.30±0.01	0.37±0.01	0.44±0.02	0.35±0.02	0.44±0.01
% Ash (MFB)[c]	13.71±0.18	14.74±0.28	19.36±0.82	15.12±0.56	20.46±0.24
Hydrolyzed Protein					
Final pH	8.60	8.61	8.65	8.75	8.70
% Total Solids	6.90±0.00	7.03±0.01	7.27±0.01	7.46±0.01	8.36±0.01
% Total Ash	0.36±0.00	0.37±0.00	0.43±0.01	0.41±0.01	0.43±0.01
% Ash (MFB)[c]	5.27±0.02	5.29±0.04	5.97±0.10	5.51±0.12	5.12±0.07

[a] Isolated from treatment of chrome shavings, sample B (cf. Table VIII).
[b] N = 3 analyses.
[c] Moisture-free basis.
Reproduced with permission from reference 56. Copyright 1992 Journal of the American Leather Chemists Association.

The effect that the chemical composition of the chrome substrates would have on the isolated protein products and chrome cakes has been described (42-43). Table VIII shows the effect of a higher ash content in the chrome shavings on the

Table VIII. Effect of Ash Content of Chrome Shavings on Character of Gelable Protein and Hydrolyzed Protein Products

Parameter [a]	Sample A	Sample B
% Moisture	52.03 ± 0.28	53.75 ± 0.16
% Ash	3.40 ± 0.12	4.24 ± 0.02
% Ash (MFB)[b]	7.09 ± 0.19	9.17 ± 0.04
Gelable Protein		
Final pH	8.94	8.90
% Total Solids	2.12 ± 0.01	2.20 ± 0.02
% Total Ash	0.23 ± 0.01	0.30 ± 0.01
% Ash (MFB)[b]	10.94 ± 0.14	13.71 ± 0.18
Hydrolyzed Protein		
Final pH	8.80	8.60
% Total Solids	5.78 ± 0.01	6.90 ± 0.00
% Total Ash	0.22 ± 0.00	0.36 ± 0.00
% Ash (MFB)[b]	3.86 ± 0.03	5.27 ± 0.02

[a] N = 3 analyses.
[b] Moisture-free basis.
[c] Chrome shavings (A & B) were treated with 6% MgO.
Reproduced with permission from reference 56. Copyright 1992 Journal of the American Leather Chemists Association.

total ash in the gelable protein and hydrolyzed protein samples. In this experiment, only magnesium oxide was used to treat the two shavings samples.

It has been reported *(48)* that the ash content of a good gel should be between 0-3%. The ash content of the samples ranged from 8.9-21% (MFB). As shown in a previous publication, these ashes contain magnesium and calcium ions as well as the more soluble sodium and potassium salts *(51-54)*. As has been shown above, this ash content reflects not only the ash content of the original shavings, but also the type of alkali used to extract the gel. Typically, in commercial gelatin preparation, the solutions are passed through ion-exchange columns to lower the ash content. The gelable solutions were passed through mixed bed ion exchange columns of two different compositions. Table IX shows the results. The first column gives the ash of the original sample and the second column gives ash contents of the deionized sample. These treated samples of gelable proteins are within the criteria set for gelatin. Both resins worked equally well. A cation exchanger may be sufficient, for it is tedious to regenerate the mixed beds; one must separate the two resins, which though not difficult is time consuming.

Table IX. Reduction of Ash Content by Mixed Bed Ion-Exchange Resins

Resin	Percent Ash[a]	
	Before	After
Amberlite MB-1		
Sample 1	12.77	0.50
Bio-Rad AG 501-X8 (D)		
Sample 1	12.77	0.43
Sample 2	17.48	0.40

[a] Moisture-free basis (MFB).
Reproduced with permission from reference 56. Copyright 1992 Journal of the American Leather Chemists Association.

The chrome cake may be treated chemically to give a recyclable chrome product, using a reported method *(55)* in which the chrome cake is dissolved in acid and the extraneous materials are eliminated by precipitation with base. Table X reports the percent residue that remains after the chrome cake is chemically treated. Samples (a) through (e) were also analyzed for the percent non-chrome insoluble ash. The low percent ash indicates that the bulk of the residue is organic, i.e. unextracted protein and/or the resins that are used in the high exhaust chrome tannages.

Table X. Characterization of Residue from Treatment of Chrome Cake

Sample	Final pH	% Residue[a]	% Ash[b]
a	1.84	9.13	0.37
b	1.85	6.33	0.18
c	1.84	7.33	0.28
d	1.85	7.46	0.25
e	1.85	10.14	0.23

[a] % residue in chrome cake, moisture-free basis (MFB).
[b] % insoluble ash in chrome cake (MFB).
Reproduced with permission from reference 56. Copyright 1992 Journal of the American Leather Chemists Association.

A cost and return estimate of the described treatments has been calculated. The cost of chemicals, energy, labor, and equipment and the return from chrome cake, savings on landfilling, and the recovered protein were factored into the equation. It was found that the tanner should realize a $1.77 return from the total chrome-containing solid waste generated from each cattlehide, when using the technology of the newer two-step process and recycling the enzyme-containing solution twice. The new treatment is more profitable and this is influenced by the return on the gelable protein. Recycling the enzyme will increase profits slightly in the two-step treatment (from $1.61 to $1.77), but will definitely improve the cost effectiveness of the original one-step treatment (from a loss of $0.17 to a profit of $0.11). These savings are mainly the result of lowered costs of evaporation.

In conclusion, high quality gelable protein and hydrolyzed protein products can be isolated from chromium-containing leather waste. Depending on choice of alkali, the process can be varied to give a desired end product, such as molecular weight distribution and Bloom value. It has been shown that the choice of alkali for treatment of chromium-containing waste influences the chemical composition of the isolated protein products. The chemical composition of the original chromium waste product also contributes to the chemical makeup of the protein products. A higher percentage of the ash is extracted with the gelable protein, and if this ash is too high for the desired end product, it can be removed by ion-exchange resins. This study has also shown that a variety of alkalinity-inducing agents can be used to treat the waste, depending on the desired composition of the end product or compatibility with the chemicals used in chrome recycling in the tannery system. It has also been shown that the chrome cake isolated in these treatments can be chemically treated to remove undissolved protein or the resins used in the high exhaust chrome tannages, so that a recyclable product can be recovered. A cost estimate has been calculated and indicated that a profit can be achieved if the new two-step process is run and the enzyme is recycled.

Reference to a brand or firm name does not constitute endorsement by the U.S. Department of Agriculture over others of a similar nature not mentioned.

Literature Cited

1.   Holloway, D.F., U.S. Patent 4,100,154 (1978).
2.   Guardini, G., U.S. Patent 4,483,829 (1983).
3.   Galatik, A., Duda, J. and Minarik, L., Czech. CS 252,382 (1988).
4.   Przytulski, S. and Supera, A. *Przegl. Skorzany* 1980, *35*, 306-312.
5.   Farbenfabriken Bayer A.-G., Fr. Demande 2,004,440 (1969).
6.   Toshev, K., Pesheva, M. and Milyanova, K. *Kosh. Obuvna Prom-st.* 1980, *21*, 8-10.
7.   Ohtsuka, K., Japan Patent 73 29,145 (1973).
8.   Bezak, A., Matyasovsky, J., Henselova, M. and Varkonda, S. *Kozarstvi* 1989, *39*, 20-23.

9. Fujimoto, Y. and Teranishi, M., Japan Patent 74 20,102 (1974).
10. Trakhtenberg, S.I. and Korostyleva, R.N., *Kozh.-Obuvn. Prom-st.,* **1982,** *24,* 40-41.
11. Okamura, H. and Shirai, K. *J. Am. Leather Chem. Assoc.,* **1976,** *71,* 173-179.
12. Cot, J. and Gratacos, E., *AQEIC Bol. Tec.,* **1975,** *26,* 353-376.
13. Jones, B.H., U.S. Patent 4,086,319 (1978).
14. Ioan, F., Filofteia, D. Rom. Patent 65,477 (1978).
15. Okamura, H., Tanaka, N. and Yashura, K. *Hikaku Kagaku,* **1981,** *27,* 83-88.
16. Cioca, G. and Fertell, P.A., Fr. Demande 2,495,054 (1982).
17. Klasek, A., Kaszonyiova, A. and Sykorova, M., *Kozarstvi,* **1985,** *35,* 102-104.
18. Novotny, V., Rehak, P., Zurek, M. and Markova, J., Czech CS 249,803 (1988).
19. Magyar, G., Sali, G., Bertalan, Z., Juhasz, L., Zailinszky, Z. and Pakozd, G., Hung. Teljes HU 44,588 (1988).
20. Pilawski, S., Supera, A., Przytulski, S., Bienkiewicz, K., Klemm, P., and Sculc, J., Pol. Pl 115,367 (1982).
21. Alpuente Hernandis, M.L., Span. ES 543,395 (1987).
22. Stepan, A.H., Perkins, R.J. and Griggs, A.L., Ger. Offen. 2,348,229 (1974).
23. Simoncini, A., Del Pezzo, L., Grasso, G. and Bufalo, G., *Cuoio, Pelli, Mater. Concianti,* **1983,** *59,* 636-652.
24. Smith, L.R. and Donovan, R.G. *J. Am. Leather Chem. Assoc.* **1982,** *77,* 301-306.
25. Cot, J., Aramon, C., Baucells, M., Lacort, G., and Roura, M. J. Soc. Leather Technol. Chem. **1986,** *70,* 69-76.
26. Okamura, H., Ger. 1,494,740 (1973).
27. Pilawski, S. and Paryska, K. *Pr. Inst. Przem. Skorzanego* **1975,** *19,* 245-250.
28. Suseela, K., Parvathi, M.S., Nandy, S.C. and Nayudamma, Y. *Leder* **1983,** *34,* 82-85.
29. Parvathi, M.S. and Nandy, S.C. *Leather Sci.* **1984,** *31,* 236-240.
30. Suseela, K., Parvathi, M.S. and Nandy, S.C. *Leder* **1986,** *37,* 45-47.
31. Parvathi, M.S., Suseela, K. and Nandy, S.C. *Leather Sci.* **1986,** *33,* 8-11.
32. Parvathi, M.S., Suseela, K. and Nandy, S.C. *Leather Sci.* **1986,** *33,* 303-307.
33. Monsheimer, R. and Pfleiderer E., Ger. Offen. 2,643,012 (1978).
34. Hafner, B., Sommerfeld, E. and Rockstroh, B., Ger. (East) DD 212,983 (1984).
35. Hafner, B., Rockstroh, B., Sommerfeld, E., Neumann, R. and Ingeborg, A., Ger. (East) DD 243,715 (1987).
36. Taylor, M.M., Diefendorf, E.J. and Na, G.C., *Proceedings of the XXth Congress of the International Union of Leather Technologists and Chemists Societies,* October 15-19, 1989, Philadelphia, Pennsylvania.

37. Taylor, M.M., Diefendorf, E.J., Na, G.C. and Marmer, W.N., U.S. Patent 5,094,946 (1992).
38. Taylor, M.M., Diefendorf, E.J. and Na, G.C. *The Leather Manufacturer* **1990**, *108*, 10-15.
39. Taylor, M.M., Diefendorf, E.J. and Na, G.C. *J. Am. Leather Chem. Assoc.* **1990**, *85*, 264-275.
40. Taylor, M.M., Diefendorf, E.J. and Marmer, W.N. *J. Am. Leather Chem. Assoc.* **1991**, *86*, 199-208.
41. Taylor, M.M., Diefendorf, E.J. and Marmer, W.N. *Compendium of Advanced Topics on Leather Technology, XXIst Congress of the International Union of Leather Technologists and Chemists Societies,* September 23-29, 1991, Barcelona, Spain, Vol. 2, 689-703.
42. Taylor, M.M., Diefendorf, E.J., Brown, E.M. and Marmer, W.N. *Proceedings of Environmental Chemistry, Inc.,* 204th National Meeting, American Chemical Society, Washington, DC, August 23-28, 1992, 26-28.
43. Taylor, M.M., Diefendorf, E.J., Brown, E.M. and Marmer, W.N. *J. Am. Leather Chem. Assoc.* **1992**, *87*, 380-388.
44. Taylor, M.M., Diefendorf, E.J., Brown, E.M. and Marmer, W.N., U.S. Patent 5,271,912 (1993).
45. Taylor, M.M., Diefendorf, E.J., Phillips, J.G., Feairheller, S.H. and Bailey, D. G. *J. Am. Leather Chemists Assoc.* **1986**, *81*, 4-18.
46. Laemmli, U.K. *Nature* **1970**, *227*, 680-685.
47. AOAC Method 948.21.
48. Rose, P.I. "Inedible gelatin and glue" *Inedible Meat By-Products,* Advances in Meat Research, Vol. 8, (A.M. Pearson and T.R. Dutson, Eds.) Elsevier Applied Science, London and New York, 1992, pp. 217-263.
49. Hauck, R.A. *J. Am. Leather Chemists Assoc.* **1972**, *67*, 422-430.
50. Langerwerf, J.S.A. and de Wijs, J.C. *Das Leder* **1977**, *28*, 1-8.
51. Taylor, M.M., Diefendorf, E.J., Artymyshyn, B., Hannigan, M.V., Phillips, J.G., Feairheller, S.H. and Bailey, D.G. The chemical and physical analysis of contemporary thru-blue tanning technology. 1984. (A Compendium of analytical data distributed to the tanning industry. Can be obtained from HLW Research Unit, ERRC.)
52. Taylor, M.M., Diefendorf, E.J., Phillips, J.G., Hannigan, M.V., Artymyshyn, B., Feairheller, S.H. and Bailey, D.G. *J. Am. Leather Chemists Assoc.* **1986**, *81*, 19-34.
53. Taylor, M.M., Diefendorf, E.J., Hannigan, M.V., Artymyshyn, B., Phillips, J.G., Feairheller, S.H. and Bailey, D.G. *J. Am. Leather Chemists Assoc.* **1986**, *81*, 43-61.
54. Taylor, M.M., Diefendorf, E.J., Civitillo Sweeney, P.M., Feairheller, S.H. and Bailey, D.G. *J. Am. Leather Chemists Assoc.* **1988**, *83*, 35-45.
55. Okamoto, Y. and Katano, S., Japan Patent 74 16,358 (1974).
56. Taylor, M.M., Diefendorf, E.J., Thompson, C.J., Brown, E.M. and Marmer, W.N. *J. Am. Leather Chem. Assoc.,* in press (1994).

RECEIVED May 24, 1994

Chapter 13

Embrittlement and Rejuvenation of Bacterial Poly[(R)-3-hydroxybutyrate]

G. J. M. de Koning[1]

Centre for Polymers and Composites (CPC), Eindhoven University of Technology, P.O. Box 513, 5600 MB Eindhoven, Netherlands

A major drawback of bacterial poly[(R)-3-hydroxybutyrate] (PHB) is its intrinsic brittleness. Although as-moulded PHB shows ductile behaviour, upon storage at ambient temperature a detrimental ageing process embrittles the material and limits its application possibilities. This remarkable embrittlement is delineated in the present study and could be attributed to progressive crystallization. Up to now, most attempts to improve the mechanical properties of PHB have focussed on incorporating comonomers, such as 3-hydroxyvalerate, but then the mechanical improvements are at the expense of the production costs. Alternatively, we found that by using a simple annealing treatment PHB homopolymer can be toughened while subsequent ageing is prevented to a large extent.

The expansion of plastic (polymer) production over the past few decades has caused considerable waste problems, since these synthetic materials do not degrade easily and hence accumulate in the environment. The annual disposal of over 10 million tons of plastics in both the US and EC (*1*) has raised the demand for degradable substitutes to reduce the environmental impact of polymer materials. Several types of biodegradable polymers have emerged. Some of these could well become obsolete when more rigorous standards for biodegradability are defined and applied (*2*). One polymer material of which the biodegradability has always been beyond dispute (*3,4*) is exemplified by poly[(*R*)-3-hydroxybutyrate] (PHB), a biopolyester accumulated by a wide variety of bacteria as a reserve of carbon and energy (*5*). Due to its enzymatic synthesis, this biopolymer has an exceptional stereochemical regularity. The polymer is completely isotactic (*6*) and, as a consequence, PHB is capable of crystallizing.

[1]Current address: Institut für Biotechnologie, Eidgenössische Technische Hochschule, ETH Hönggerberg (HPT), CH–8093 Zurich, Switzerland

Crystallinities are typically in the range of 55 - 80% (*7*). The polymer shows a glass transition temperature (T_g) at approximately 5 °C and a crystalline melting point (T_m) around 175 °C (*7-9*). Importantly, its hydrophobicity is unique among natural polymers and makes PHB excel its biodegradable competitors in moisture resistance.

Industrial interest in PHB has flagged, however, because of two major drawbacks. First, PHB has a poor melt stability, though melt-processing of PHB certainly is not impracticable provided that both melt temperature and melt residence time are minimized. Second, PHB is brittle. Although as-moulded PHB shows ductile behaviour, upon storage at ambient temperature a detrimental ageing process embrittles the material.

Up to now, most attempts to overcome the intrinsic (*10*) brittleness of PHB have focussed on incorporating comonomers. For example, PHB copolymers with 3-hydroxyvalerate (HV) have been commercialised by Zeneca under the trademark Biopol. These copolymers are obtained by using specific additives in the growth medium of the bacteria. Like PHB, the PHB/HV-copolymers exhibit the embrittlement phenomenon as well (*11*), but the ultimate maximum elongation is somewhat higher (*7*). Moreover, the melting point of PHB drops considerably by incorporating HV units (*12*), thus widening the processability window. Notwithstanding these superior properties, copolymer production costs are higher than for PHB homopolymer. The required additives are expensive and, owing to their toxicity in the culture, production yields are somewhat lower (*12*) than for PHB homopolymer. Moreover, the presence of comonomer affects the crystallization kinetics of the polymer (*13-15*), giving rise to longer processing cycle times.

It is of major importance to overcome the brittleness of PHB, since it seriously limits its applicability as a biodegradable plastic. Since PHB/HV copolymers form an moderately successful approach, the objective of the present study was to improve the toughness of PHB homopolymer. A logical first step was to attempt to delineate the mechanism responsible for the remarkable embrittlement of PHB.

Experimental

Materials.   The PHB homopolymer investigated was a Biopol G08 sample (M_w=539 kg·mol^{-1}, M_w/M_n=3.5, T_m=174 °C), provided by Zeneca Bioproducts (Billingham, UK). The powder was mixed with 1 wt% boron nitride in a Hobart mixer for approximately 10 min. Boron nitride acts as a nucleating agent (*16*) to enhance the otherwise slow crystallization process. Moreover, the presence of a consistent concentration of nucleating agent is essential in order to eliminate any variation of the spherulite size and concomitant mechanical effects.

Sample Preparation.   The powder mixture was extruded into a single strand of 4 mm diameter using a Betol 2520 extruder (diameter 25 mm; filters 60 and 100 mesh) operated at a maximum of 180 °C and at a screw speed of 100 rev·min^{-1}. The strand was crystallized in a 60 °C water bath and pelletized. After drying at 40 °C for 20 h, the pellets were injection-moulded using a Boy 15S injection-moulding machine. Processing specifications: maximum barrel temperature 180 °C; injection time 15 s; screw speed 220 rev·min^{-1}; injection pressure 5 MPa; mould temperature 60 °C; cooling time 15 s. Ageing of the resulting specimens was established by storage at ambient conditions for a specified time. After 800 days of storage some samples were annealed in a Heraeus hot-air oven.

Mechanical Testing. Tensile testing was performed at temperatures ranging from 20 to 60 °C using an Instron 1122 fitted with a Nene data analysis system. An initial clamp separation of 50 mm and a crosshead speed of 20 mm·min^{-1} were used. The injection-moulded specimens were dumbbell-shaped according to ISO R 537/2, their prismatic part measuring 40 x 5.3 x 1.9 mm.

The Izod impact strength was determined using a Zwick pendulum impact testing machine on injection-moulded specimens with dimensions of 5 x 13 x 50 mm, which were provided with a V-notch possessing a tip radius of 1 mm. All tests were conducted under ambient conditions.

Dynamic mechanical measurements were performed with a Polymer Laboratories Dynamic Mechanical Analyser, operated in the single cantilever bending mode. Injection-moulded specimens (1.9 x 5.3 x 12 mm) were investigated at a measuring frequency of 1 Hz and a heating rate of 2 °C·min^{-1}.

Dilatometry. Dilatometry was conducted in a Hg-dilatometer (*17*). The dilatometer was packed with PHB pellets and mercury was added under vacuum. The sealed dilatometer was subjected to the temperature profile of the processing cycle and, subsequently, positioned in a thermostatically controlled water bath. The density of the sample was monitored over a period of one month. The measured densities were converted into sample densities at 20 °C using a thermal expansion coefficient of 25×10^{-5} °C^{-1}. The density of PHB, ρ_{PHB}, was obtained by correcting the sample density for the presence of 1 wt% boron nitride (BN):

$$\rho_{PHB} = 0.99 \cdot \frac{\rho_{sample}}{1 - 0.01 \cdot \frac{\rho_{sample}}{\rho_{BN}}} \qquad where \; \rho_{BN} = 2.25 \; g \cdot cm^{-3}$$

Densities could be translated into the volume fraction and the mass fraction crystallinities, X_v and X_m respectively, using:

$$X_v = \frac{\rho_{PHB} - \rho_a}{\rho_{cr} - \rho_a} \quad ; \quad X_m = X_v \cdot \frac{\rho_{cr}}{\rho_{PHB}}$$

with   density of amorphous phase (*8*): $\rho_a = 1.179$ g·cm^{-3}
density of crystalline phase (*8*): $\rho_{cr} = 1.279$ g·cm^{-3}

Microscopy. Pieces of injection-moulded samples were trimmed ready for microtoming. Thin sections were obtained by ultramicrotomy at ambient temperature using a Reichert Ultracut E Microtome. The thin sections (thickness: 100 nm) were stained in ruthenium tetroxide vapour for 2 hrs at room temperature. Transmission electron microscopy was performed using a Jeol JEM 2000 FX microscope operated at 80 kV. The spherulitic texture of a slice of a PHB-pellet was monitored using a Zeiss optical microscope equipped with polarizing filters.

Differential Scanning Calorimetry (DSC). Melting endotherms were recorded using a Perkin-Elmer DSC-7 differential scanning calorimeter at a heating rate of 20 °C·min^{-1}. Indium was used for temperature and heat of fusion calibration. Mass fraction crystallinities could be calculated from the melting peak area, assuming that the heat of fusion of 100% crystalline PHB (8) amounts to 146 J·g^{-1}.

X-ray Scattering. In order to determine the lattice parameters, wide-angle X-ray scattering (WAXS) patterns of injection-moulded samples (thickness: 1.9 mm) were obtained using a Philips PW 1729 X-ray Generator (Ni-filtered CuK$_\alpha$) and a Statton camera. Small-angle X-ray scattering (SAXS) measurements were conducted at ambient temperature with a Rigaku type Kratky camera using infinite slit geometry. Ni-filtered CuK$_\alpha$ radiation was produced using a Rigaku rotating anode device operated at 50 kV and 150 mA. The sample to detector distance was 280 mm. Each sample had a thickness of 1.9 mm and was exposed for 30 min. Scattering patterns were collected with a linear position sensitive detector (Braun OED-SOM). After correction for the empty cell scattering and subtraction of the background scattering, the smeared intensity was obtained. Subsequent desmearing yielded the corresponding pinhole collimation scattering pattern. The periodicity L was then found as the reciprocal of the scattering maximum in a plot of the Lorentz corrected intensity.

Results and Discussion

Effects of Storage on Mechanical Properties. Figure 1 presents the stress-strain behaviour of moulded PHB after different periods of storage at ambient conditions. Initially, the material showed a ductile behaviour with a maximum elongation of approximately 50%. After two weeks of storage, however, the maximum elongation had dropped below 10% reflecting the embrittlement of the material. An ultimate level of 5% was reached at ca. 100 days of storage, after which no further deterioration was observed. Figure 2 shows the corresponding changes in tensile modulus and Izod impact strength. Both quantities show a logarithmic change with time and deterioration ceased after some 100 days.

Figure 3 presents the dynamic mechanical characteristics as a function of storage time. The drop in storage modulus at the glass transition decreased and eventually disappeared and the dynamic loss peak showed a drastic reduction. Both these trends reflect a loss of the relaxation strength, i.e. the pliability, of the amorphous phase. In a creep experiment, this relaxation strength is defined as the normalized difference in strain between the initial unrelaxed state and the final relaxed state. For the case of a standard linear solid, it can be demonstrated that both the drop in storage modulus and the dynamic loss peak maximum (tan δ_{max}) scale with this relaxation strength (*18*). Tan δ_{max} is closest to the original definition of relaxation strength in being normalized to a dimensionless quantity. Also in real polymer systems, tan δ_{max} is an appropriate, though not proportional measure of the relaxation strength.

When tan δ_{max} was monitored during storage of PHB, this quantity showed a logarithmic decrease with time until it stabilized at ca. 100 days (Figure 4). The striking parallel between the data from Figures 2 and 4 strongly suggests a close connection and possibly a causality between the relaxation strength of the amorphous phase and the macroscopic mechanical properties.

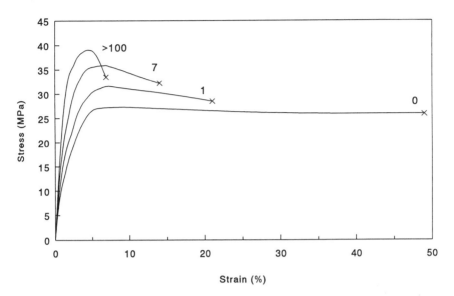

Figure 1. Stress-strain curves under ambient conditions for moulded PHB samples after different periods of storage. The storage times, in days, are indicated in the curves. (Adapted from ref. 28.)

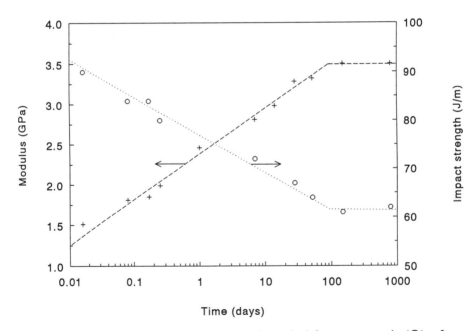

Figure 2. Tensile modulus (+) and Izod notched impact strength (O) of moulded PHB versus storage time. (Reproduced from ref. 28 with permission of the publishers Butterworth Heinemann Ltd.©)

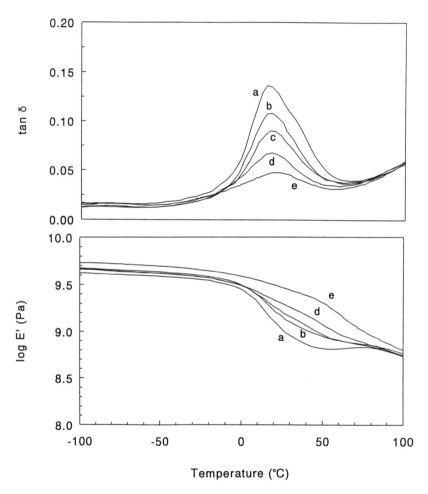

Figure 3. Dynamic mechanical spectra of moulded PHB in the region encompassing the glass transition after storage for (a) 0 h, (b) 2 h, (c) 1 day, (d) 8 days, (e) 150 days. (Adapted from ref. 28.)

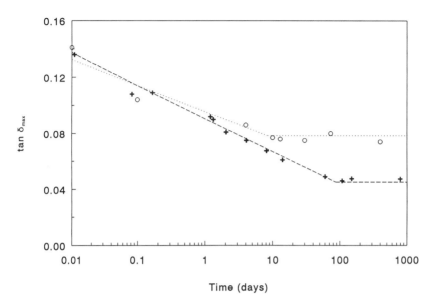

Figure 4. Dynamic loss peak maximum vs. time of storage at room temperature for as-moulded PHB (+) and for PHB annealed for 10 min at 147 °C (O).

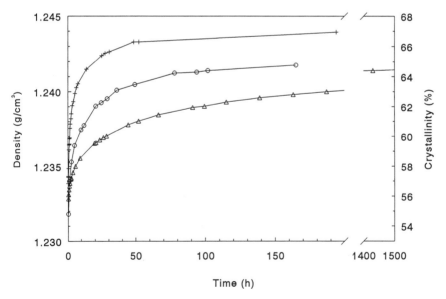

Figure 5. Density changes in as-moulded PHB as observed by dilatometry at 25 °C (Δ), 43 °C (O), and 60 °C (+). (Adapted from ref. 28.)

Physical Ageing versus Progressive Crystallization. Two processes might be involved in the embrittlement of PHB. First, PHB might be prone to 'physical ageing', as was claimed before by Scandola and colleagues (*11*). Second, moulding for 15 s at 60 °C may not be sufficient to complete the crystallization process, so that progressive crystallization will occur upon subsequent storage at room temperature. The observed trends in modulus, impact behaviour and dynamic mechanical characteristics are consistent with either mechanism.

Physical Ageing. Physical ageing is a phenomenon characteristic of all glassy materials (*19*) and in semicrystalline polymers it is known to persist at temperatures slightly higher than T_g (*20-22*). Owing to the non-equilibrium character of the glassy state, the residual mobility tends to produce molecular rearrangements that drive the free volume closer to its equilibrium value. The resulting volume relaxation restricts the mobility, which affects all related properties. This process can be completely reversed by regenerating free volume, which can be achieved by heating the material above T_g. Since this treatment does not affect the crystals, its reversible character enables physical ageing to be distinguished from progressive crystallization. Tensile testing demonstrated that elevating the temperature up to 60 °C had no effect at all on the maximum elongation of PHB. Since this temperature is well above T_g, this observation eliminates vitrification and physical ageing as possible causes of the embrittlement of PHB.

Progressive Crystallization. Dilatometry (Figure 5) showed that the storage of PHB is accompanied by a significant increase in density. If this density increase is attributable entirely to crystallization, the initial 56% crystallinity of as-moulded PHB would increase by an additional seven percent upon 200 h storage at 25 °C. Similarly, from the DSC melting endotherm it was calculated that the crystallinity increased from 55 to 61%. Although the DSC data are less accurate ($\pm 3\%$ compared to $\pm 0.2\%$ for dilatometry), their resemblance with the results from dilatometry indicates that at least the greater part of the density increase derives from crystallization. After a period of two months, dilatometry revealed a crystallinity of 65%, indicating that crystallization progressed over a considerable time parallel to the mechanical changes.

Secondary Crystallization. Using WAXS, no significant variations were observed as a function of time in the lattice parameters. The gain in crystallinity upon storage apparently does not result from some crystal phase transition. Neither does it result from extended spherulitic growth: optical microscopy revealed a full-grown and invariable spherulitic texture throughout the ageing process, which justifies the conclusion that primary crystallization was completed. However, when crystallizing from the melt, a polymer does not usually reach its ultimate crystallinity in a single step. Even after spherulites impinge, the crystallinity often continues to rise more slowly (*23*). Progressive crystallization of PHB is likely to consist of this 'secondary' crystallization. Figure 5 shows that even at 60 °C complete crystallization required at least two days. Obviously, secondary crystallization cannot be completed within the processing cycle. Upon subsequent storage at room temperature, the close proximity of T_g hinders the diffusion of chains and, therefore, seriously limits the rate of any crystallization process.

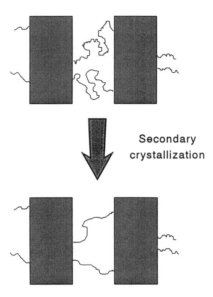

Figure 6. Schematic visualization of the constraining effect of secondary crystallization on the amorphous phase.

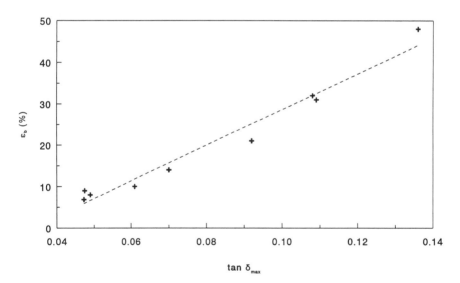

Figure 7. Elongation at break (ϵ_b) versus the loss peak maximum (tan δ_{max}) during ageing of moulded PHB.

During secondary crystallization of PHB the crystallinity can increase by different processes (*24*). Since SAXS measurements did not reveal any increase in scattering power, formation of new lamellar stacks is highly unlikely. Furthermore, the periodicity did not vary over a period of three months ruling out the possibility of new crystals being formed within lamellar stacks. The apparent certainty, that the morphology is constant throughout the ageing process, only leaves the possibility of crystal perfection. Crystal defects might be located within the lamellar core and amending should give rise to a sharper WAXS pattern. However, the changes in the WAXS pattern upon ageing were too small to justify any conclusion. Defects might also be located in the crystalline-amorphous interface, rectification of which is more likely and would give rise to thickening of the lamellar core. However, SAXS could not furnish any evidence of isothermal lamellar thickening, since the invariant is highly insensitive for small changes within the crystallinity range from 30 to 70%.

Constraint from Secondary Crystallization. The relatively small loss of amorphous fraction upon secondary crystallization is attended by drastic mechanical effects, e.g. virtually complete loss of the relaxation strength of the amorphous phase. Such a reduction of relaxation strength can be caused either by a decreasing amorphous fraction or by increasing constraints in the amorphous regions. The latter obviously plays a dominant role in the case of PHB. Apparently, secondary crystallization tightly constrains the amorphous phase between the crystals. Consistently, whereas the as-moulded PHB readily dissolved in chloroform, the aged material required extensive refluxing indicative of a reduced accessability of the amorphous phase. Whatever mechanism is invoked for crystal perfection, all can be argued to constrain the amorphous chains. This is schematically visualized in Figure 6. The constrained amorphous phase can be compared to a pre-strained rubber. The constraint obviously will increase its modulus and reduce its relaxation strength.

Mechanism of Deformation. The lower elastic modulus of as-moulded material (Figure 1) is a direct consequence of the higher relaxation strength of its amorphous phase, since generally the amorphous phase almost entirely accounts for the elastic part of the deformation (*25*) between T_g and T_m. However, Figure 1 demonstrates that it is the ability to undergo plastic deformation which is responsible for the toughness of as-moulded PHB. Semicrystalline polymers generally (*25*) deform plastically by slip in the crystals. However, polymer crystals rarely possess the five independent slip systems necessary for a general change of their shape, as must occur in the deformation of a polycrystalline aggregate. In order to enable plastic deformation, it is therefore essential that the amorphous regions between adjacent crystals allow a certain amount of adjustment to accommodate some rotation of the crystals. Apparently, upon ageing of PHB an increasing fraction of the amorphous regions becomes too constrained to allow rotation of the adjacent crystals, giving rise to a gradual decrease in the macroscopic strain at break. This is consistent with the observed parallel between $\tan \delta_{max}$, representing the relaxation strength of the amorphous phase, and the macroscopic strain at break (ϵ_b) (Figure 7).

PHB as Unique. The changes in properties resulting from ageing, as reported in literature (*19-22*) for other semicrystalline polymers, are only small compared to the drastic embrittlement which occurs upon ageing of as-moulded PHB. We believe that this peculiarity is related to the unique morphology of PHB. Owing to its low nucleation density, PHB can be crystallized isothermally at large supercoolings and hence unusually fine morphologies can be achieved. For example, when crystallized below 60 °C, PHB formed thin lamellar crystals possessing a core thickness of only 30 Å. Several other semicrystalline polymers, such as linear low-density polyethylene (26), are also known to be capable of forming thin lamellae but always with a commensurate reduction of their crystallinity. This is not the case for PHB, which developed similar crystallinities at all supercoolings. Implicitly, as-moulded PHB possesses an exceptionally large crystalline-amorphous interface per unit of volume. Any process involving this interface, such as perfection and thus ageing, will therefore be greatly pronounced compared to other semicrystalline polymers.

Fighting the Embrittlement. The above suggests that the problem concerning the embrittlement of PHB might be solved simply by creating a coarser lamellar morphology. This should be feasible by using a higher crystallization temperature (T_c). Indeed, higher lamellar thicknesses were obtained by crystallizing PHB at smaller supercoolings. However, the crystallization rate of PHB above 120 °C is almost negligible (13). Therefore, long crystallization times were required giving rise to considerable degradation.

Using DSC, a peak melting temperature of ca. 174 °C was observed for all samples independent of T_c, which is completely inconsistent with the established fact that the melting point increases with the lamellar thickness. Obviously, the peak maximum of the melting endotherm does not represent the melting temperature of the original crystals. Recent studies (13-15) related this observation to a rearrangement of the initial crystal morphology. Upon heating, the crystals experience a continuous melting and structural reorganization into thicker and more stable crystals, which therefore have higher melting temperatures. The peak temperature only indicates the point at which the net difference between melting and recrystallization passes through a maximum. In contrast to crystallization directly from the melt, recrystallization above 120 °C is rapid because annealing produces a well nucleated 'melt' due to the abundance of intact crystals and crystal residues. Therefore, the favourable coarser morphology may be obtained without degradation by annealing at high temperatures.

Morphological Changes upon Annealing.   Figure 8 presents the crystal morphologies of PHB before and after annealing at 150 °C as viewed by TEM. It clearly shows that the lamellar crystals are grouped in stacks and that annealing increased the lamellar thickness. The lamellar texture of the original material was found to be exceptionally fine compared to other semicrystalline polymers. Annealing produced a much coarser lamellar texture comparable to that of polypropylene. Consistent results were obtained from the SAXS measurements. The initial periodicity, which was as small as 70 Å, almost doubled upon annealing at 150 °C. Surprisingly, both dilatometry and DSC showed that the increase in lamellar thickness was accompanied with only a slight increase in crystallinity. This implies that the total number of crystals was reduced, which is consistent with the concept of melting and recrystallization.

As annealing almost doubles the periodicity, the crystalline-amorphous interface per unit of volume, which is proportional to the reciprocal periodicity, may decrease by some factor two. As described previously, the constraining of the amorphous phase upon ageing should be attributed to secondary crystallization, more specifically crystal perfection, which may involve rectifying the chain conformation and thickening of the crystal core. The latter simply consists of perfection processes in the crystalline-amorphous interface at the expense of the amorphous layer, e.g. the rearrangement of loops in the fold region. The resulting constraint imposed on the amorphous regions should therefore be lower in annealed material, since annealing reduces the ratio of interface area to amorphous material.

Toughening by Annealing. The ageing of as-annealed PHB was examined in an analogous way to the embrittlement of the as-moulded material. Dilatometry of the annealed material showed that storage at 25 °C elevated the density slightly. Notably, the volume of material involved in crystal perfection, i.e. the total amount of interface material, decreased upon annealing. Consistently, the increase in density (crystallinity) upon ageing in as-moulded material was much higher than that in annealed material.

Both as-moulded and as-annealed PHB initially showed similar high relaxation strengths (Figure 4), i.e. their amorphous phase is 'loose'. However, as the extent of subsequent secondary crystallization was smaller in annealed material, so was the resulting constraint inmposed on the amorphous phase. After annealing, tan δ_{max} initially decreased similarly as in as-moulded material but then stabilized at a level twice as high and therefore in an earlier stage. Accordingly, annealing restored the original toughness of the material and subsequent deterioration was minor and lasted for only a few days (Figure 9). Annealing at 150 °C increased the ultimate elongation at break from 5% to over 25%.

Conclusion

The annealing treatment presented has improved the fracture behaviour of PHB homopolymer to such an extent that the scope of its application possibilities may widen significantly. PHB/HV applications, currently in the market or being developed, might now beneficially be produced from PHB homopolymer. Moreover, applications for which PHB was previously considered to be too brittle need to be reevaluated. Devices for bone fracture fixation (27), such as resorbable bone plates and splints, are now under investigation. In practice, PHB objects could be annealed by shortly storing them in an oven. Annealing of thin products, such as films and coatings, ideally could be performed by conveying them through a heat source directly after primary crystallization.

Acknowledgements

The author wishes to thank Prof. P.J. Lemstra and Prof. H.E.H. Meijer (CPC, The Netherlands) for many invaluable discussions throughout the entire study. Prof. H. Reynaers and coworkers (University of Leuven, Belgium) are gratefully acknowledged for kindly providing their SAXS equipment and software and for offering concomitant assistance and expertise. The author is indebted to the staff of Zeneca BioProducts (Billingham, UK) for helpful discussions, technical assistance and financial support.

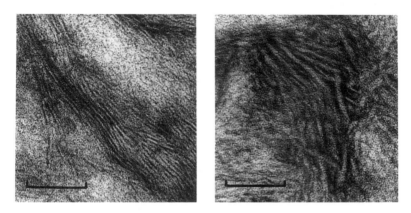

Figure 8. TEM micrographs of moulded PHB samples annealed for (a) 0 min
and (b) 10 min at 147 °C (scale bar: 1000 Å).

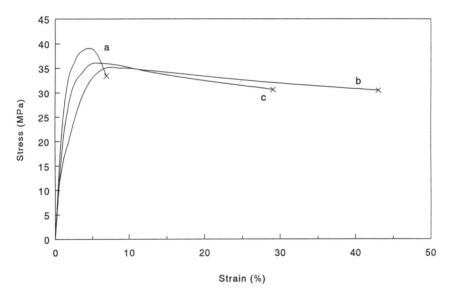

Figure 9. Stress-strain curves under ambient conditions for moulded PHB
samples which were (a) stored for 800 days, (b) subsequently annealed for 10
min at 147 °C, and (c) stored for another 50 days. (Reproduced from ref. 28
with permission of the publishers Butterworth Heinemann Ltd.©)

Literature Cited

1. Chem Systems International Ltd., *Biodegradable Plastics: Future Trends in Western Europe and the USA*; N.R.L.O.-report 89-34; N.R.L.O.: The Netherlands, 1989.
2. Sadun, A.G.; Webster, T.F.; Cooper, D.G. *Breaking down the biodegradable plastics scam*; Greenpeace report; Center for Biology of Natural Systems, Flushing: New York, 1990.
3. Krupp, L.R.; Jewell, W.J. *Environ. Sci. Technol.* **1992**, *26*, 193.
4. Barak, P.; Coquet, Y.; Halbach, T.R.; Molina, J.A.E. *J. Environ. Qual.* **1991**, *20*, 173.
5. Dawes, E.A.; Senior, P.J. *Adv. Microbiol. Physiol.* **1973**, *10*, 135.
6. Doi, Y.; Kunioka, M.; Nakamura, Y.; Soga, K. *Macromol.* **1986**, *19*, 1274.
7. Holmes, P.A. In *Development in Crystalline Polymers*; Bassett, D.C., Ed.; Elsevier: London, 1988, Vol. 2; p. 1.
8. Barham, P.J.; Keller, A.; Otun, E.L.; Holmes, P.A. *J. Mater. Sci.* **1984**, *19*, 2781.
9. Owen, A.J. *Coll. Pol. Sci.* **1985**, *263*, 799.
10. De Koning, G.J.M.; Lemstra, P.J.; Hill, D.J.T.; Carswell, T.G.; O'Donnell, J.H. *Polymer* **1992**, *33*, 3295.
11. Scandola, M.; Ceccorulli, G.; Pizzoli, M. *Makromol. Chem. Rapid. Commun.* **1989**, *10*, 47.
12. Doi, Y. et al. *Appl. Microbiol. Biotechnol.* **1988**, *28*, 330.
13. Organ, S.J.; Barham, P.J. *J. Mater. Sci.* **1991**, *26*, 1368.
14. Owen, A.J.; Heinzel, J.; Škrbić, Ž.; Divjaković, V. *Polymer* **1992**, *33*, 1563.
15. Barker, P.A. *Ph.D. Thesis*; University of Bristol: Bristol, UK, 1993.
16. Barham, P.J. *J. Mater. Sci.* **1984**, *19*, 3826.
17. Wilson, P.S.; Simha, R. *Macromolecules* **1973**, *6*, 902.
18. Ward, I.M. *Mechanical Properties of Solid Polymers*; Wiley-Interscience: London, UK, 1971; pp. 107-109.
19. Struik, L.C.E. *Physical Ageing of Amorphous Polymers and Other Materials*; Elsevier: Amsterdam, The Netherlands, 1978.
20. Struik, L.C.E. *Plast. Rub. Proc. Applic.* **1982**, *2*, 41.
21. Struik, L.C.E. *Polymer* **1987**, *28*, 1524.
22. Struik, L.C.E. *Polymer* **1987**, *28*, 1534.
23. Mandelkern, L. *Crystallization of Polymers*; McGraw-Hill: New York, 1964.
24. Zachmann, H.G.; Wutz, C. *Polym. Prep.* **1992**, *33*, 261.
25. Bowden, P.B.; Young, R.J. *J. Mater. Sci.* **1974**, *9*, 2034.
26. Defoor, F. et al. *Macromolecules* **1993**, *26*, 2575.
27. Doyle, C.; Tanner, E.T.; Bonfield, W. *Biomaterials* **1991**, *12*, 841.
28. De Koning, G.J.M.; Lemstra, P.J. *Polymer* **1993**, *34*, 4098.

RECEIVED May 24, 1994

Chapter 14

Bacterial Conversion of a Waste Stream Containing Methyl-2-hydroxyisobutyric Acid to Biodegradable Polyhydroxyalkanoate Polymers

L. P. Holowach[1], G. W. Swift[1], S. W. Wolk[1], and L. Klawiter[2]

[1]Rohm and Haas Company, Research Division, 727 Norristown Road, Spring House, PA 19477
[2]Rohm and Haas Company, Process Economics Center, Engineering Division, Route 413 & State Road, Bristol PA 19007

Bacteria were used to study the feasibility of converting a liquid organic manufacturing waste stream to solid biodegradable polyhydroxyalkanoate (PHA) polymers. Bacteria were isolated from activated sludge obtained from an acrylate manufacturing plant. Representatives of several genera, including *Bacillus*, *Pseudomonas*, and *Alcaligenes* were found to be well-adapted to growth on methyl-2-hydroxyisobutyric acid, a major component of the waste stream generated from distillation of crude methyl methacrylate. Non-adapted mixed cultures of bacteria obtained from municipal sludge and other well-studied PHA-accumulating bacteria were not able to survive in the presence of organic waste substrates under conditions where adapted bacteria grew vigorously. A preliminary economic analysis determined that the costs of using aqueous waste as a carbon feedstock was similar to the cost of using the conventional PHA production substrate, glucose. Further study of the kinetics of bacterial substrate conversion and of the feasibility of designing the physical plant for a waste conversion process are necessary to determine whether the concept of hazardous waste bioconversion to biodegradable products is a viable approach to chemical or agricultural waste management.

Microorganisms demonstrate a remarkable ability to survive in adverse environments by metabolizing a great variety of naturally-occurring and xenobiotic compounds in order to grow and multiply. This adaptability led us to explore the feasibility of using bacteria to convert organic compounds generated in organic manufacturing waste streams to solid compostable materials that could be safely disposed of in the environment. While the biotreatment industry has traditionally developed methodology to mineralize waste organic molecules, which to degrade rather than reuse organic carbon, the concept examined in this project was whether bacteria

0097–6156/94/0575–0202$08.00/0

could convert, or 'recycle', waste organic molecules to commercially useful or to environmentally innocuous products.

The desired end-product of organic waste conversion was polyhydroxyalkanoate polymers (PHAs), a class of naturally-occurring polyesters that are synthesized as storage forms of carbon and energy by a variety of bacterial genera, usually under conditions of nutritional stress (*1-5*).   PHAs are readily degraded by bacteria and fungi in terrestrial and aquatic environments. Studies with various substrates, including alkanes, alkenes, alkyl halides, aromatic, and branched-chain hydrocarbons, has resulted in the biosynthesis of novel PHAs with a diverse array of novel structures. (*6-13*).

The potential utility of PHAs in biodegradable medical and consumer products is widely recognized (*3-5*).   Biodegradability is a desirable property in products such as packaging materials, fishing nets and lines, agricultural mulches, surgical sutures, bone plates, and slow-release drug delivery systems. ICI Biological Products has developed a commercial fermentation process using *Alcaligenes eutrophus* to make copolymers of polyhydroxybutyrate and polyhydroxyvalerate (PHB/V), sold under the trade name Biopol. Biopol is used as a biodegradable packaging material for shampoo and perfume bottles.

In this study, two aspects of bioconversion were investigated. First, we examined whether bacteria adapted to living in an acrylate manufacturing waste treatment system could synthesize PHAs from acrylate substrates. In a related project, an economic analysis was performed to estimate the cost of using an aqueous manufacturing waste stream as a fermentation substrate rather than the conventional PHA substrate, glucose (*14*), to produce bacterial polyhydroxybutyrate polymers.

Selection of Manufacturing Waste Stream and Substrates for Bacterial Screening

Screening experiments were performed with the waste stream generated by the distillation of crude methyl methacrylate (MMA), obtained from the Monomers group at the Rohm and Haas Research Laboratories. This waste stream was composed of approximately 40% methyl 2-hydroxyisobutyrate (MOB) (Figure 1), the substrate selected for initial polymer production studies. Because the MMA waste was viscous and contained multiple components, including water-soluble organic molecules, polymerization inhibitors and water-insoluble components, the initial screening experiments were conducted using MOB (Aldrich Chemical Co.) as sole carbon source. Screening was also performed with a distillate of the MMA waste stream consisting of 75% methyl 2-hydroxyisobutyrate and 25% methyl 2-methoxyisobutyrate (Distillate) (Figure 1).

MOB and Distillate, typical of industrial organic wastes, required careful handling. Special safety precautions were taken when working with these substrates to protect the investigators from inhalation or skin contact with the waste stream, MOB, or Distillate. The substrates were added to autoclaved media in a chemical fume hood, rather than in a laminar flow hood. Likewise, cultures were inoculated in a chemical fume hood. To monitor for contamination introduced during this unusual inoculation procedure, uninoculated control flasks (containing waste substrates) were opened for several minutes in the chemical fume hood, then incubated in parallel with the experimental flasks to monitor for contamination introduced in the nonsterile inoculation environment. No contamination was observed to occur, presumably due to the high concentrations of methyl 2-hydroxyisobutyrate and methyl 2-methoxyisobutyrate in the experimental media. All bacterial incubations were carried out in an airtight, sealed incubator. All polymer extractions were performed in the chemical fume hood and samples were processed in tightly capped test tubes.

Sources of Bacterial Isolates

Bacteria for screening experiments were obtained from sludge taken from the trickling filter wastewater treatment facility at the Rohm and Haas Bristol, PA manufacturing plant. Sludge bacteria are known to produce a variety of PHAs (*15*), and it was hypothesized that bacteria obtained from a trickling filter system at a manufacturing site would be acclimated to continuous exposure to relatively high concentrations of a variety of small, organic molecules used in acrylate manufacture. A second sludge sample was obtained from a nearby municipal waste treatment facility. In addition, the known PHA-producing organisms *Pseudomonas oleovorans* (ATCC29347), *Alcaligenes eutrophus* (ATCC17697), and *Rhodospirillum rubrum* (ATCC25903) were cultured on MOB and Distillate to determine whether these bacteria could survive on and produce polymer from these substrates.

Selection of Bacterial Strains Using MOB as Sole Carbon Source

The process of selecting bacteria able to synthesize PHAs from MOB or Distillate was conducted in two stages. In the first stage, bacteria were selected for their ability to grow in minimal liquid media on high concentrations of MOB or Distillate as sole carbon source. One milliliter of sludge was added to 100 ml of nutrient broth (Difco) and grown overnight at 30°C with constant shaking at 150 rpm. For the initial screen, aliquots from this culture were spread onto petri plates containing minimal media (0.2% NH_4NO_3, 0.2% K_2HPO_4, 0.1% $NaHPO_4$, 0.02% $MgSO_4$, 0.01% Fe_2SO_4, 15% Difco Bacto Agar, and 0.05 to 1% (v/v) MOB or Distillate). Bacteria that grew under these conditions were transferred to media containing increasing concentrations of MOB or Distillate (up to 10% v/v). Following this initial screen for growth on MOB or Distillate, single-colony isolates were purified and sent to Microcheck, Inc (Northfield, VT), where they were identified to genus by gas chromatographic analysis of fatty acids, using a method that compared the fatty acid profile of the unknown organisms to the profiles of 7000 characterized strains in a computer database. The isolates were maintained on minimal media containing 25 mM MOB as sole carbon source.

In the second stage, the bacterial isolates were tested for the ability to synthesize PHA polymers. For this stage, a two-step fermentation process was used. In the first step, bacteria were grown overnight at 30°C, 150 rpm in nutrient broth to promote rapid cell growth. After overnight incubation, the culture was centrifuged for 5 minutes at 1500 X G, and the nutrient broth was poured off. In the second step, the cells were resuspended in the minimal media described above, except that the media contained 25 mM MOB and NH$_4$NO$_3$ was omitted to induce polymer production. The cultures were incubated for 18 to 36 hours in the sealed incubator under the same incubation conditions as used for the first step. Cells were harvested by centrifugation and lyophilized.

Polymer Characterization.

Lyophilized bacterial cells were chloroform-extracted ((7) and R.W. Lenz, personal communication) by refluxing overnight with gentle heating. The insoluble cell debris was removed by filtration through Whatman No 1 filter paper. The chloroform-soluble extract was partially dried in a rotary evaporator to reduce the volume of chloroform, then polymer was precipitated by addition of a minimal amount of cold (4°C) methanol. The polymer was pelleted by centrifugation in a clinical centrifuge, and excess liquid was poured off. The pellet was rinsed twice with methanol, then and the polymer was allowed to air dry before being redissolved in deuterated chloroform and analyzed by NMR on a Bruker AMX500 spectrometer (^{1}H = 500 MHz)

Bacterial Polymer Synthesis using MOB and Distillate as Sole Carbon Sources

A number of bacteria obtained from the industrial trickling filter sludge system were able to grow on high (up to 10% v/v) concentrations of MOB and Distillate. Table I is a partial list of genera identified from the screening process that were able to grow on MOB or Distillate as sole carbon sources. Bacteria from municipal sludge and PHA-producing strains obtained from ATCC, *Pseudomonas oleovorans*, *Rhodospirillum rubrum* and *Alcaligenes eutrophus* were unable to survive at concentrations of MOB or Distillate of greater than 0.5% to 1% in the media (v/v).

Table I. Bacterial genera found to grow on high (up to 10% v/v) concentrations of methyl-2-hydroxyisobutyrate (MOB) or Distillate. These bacteria were originally isolated from sludge obtained from the wastewater treatment plant at an acrylate manufacturing facility.

Alcaligenes
Xanthomonas
Bacillus
Clavibacter
Serratia
Pseudomonas
Enterobacter
Ochrobactrum
Erwinia

Flavobacterium
Brevibacterium

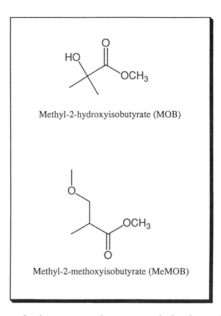

Figure 1. Structures of substrates used to screen sludge bacteria for ability to grow on components of the organic waste stream generated during distillation of methyl methacrylate.

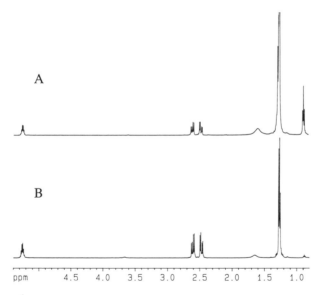

Figure 2. ^{1}H-NMR analysis of chloroform-soluble fraction obtained from *Bacillus* sp. grown on methyl 2-hydroxyisobutyric acid as sole carbon source. A. PHB isolated from *Bacillus sp.* grown on MOB as sole carbon source. B. PHB standard.

Analysis of Polymer Composition of Bacteria Cultured on MOB as Sole Carbon Source

A number of bacteria that were able to grow on MOB as sole carbon source were also able to synthesize PHA polymer under nitrogen-limiting conditions. ^{1}H-NMR analysis demonstrated that the most prevalent polymer product made was polyhydroxybutyrate (PHB) (Figure 2). NMR analysis indicated that another type of polymer was also synthesized in the presence of MOB however, the structure of this other polymer has not been determined at this time.

Hypothesized Pathway for Metabolism of MOB to PHB

The pathway for conversion of MOB to PHB is hypothesized to proceed through a pyruvate intermediate, with removal of an α-methyl group and the methyl moiety from the methoxy group of MOB (Figure 3). Pyruvate may be converted to acetyl CoA, which would allow the MOB backbone to be processed within the normal PHB synthesis pathway (5).

Economic Analysis of Chemical Waste Management Using Bacteria

In a related study, the economic feasibility of using polymer-forming bacteria to convert a dilute aqueous acetic acid waste stream to PHB/V was analyzed. The commercial production of PHB and PHB/V copolymers is currently limited to specialty applications. Because production costs are high, market prices for these biodegradable polymers are approximately five to ten dollars per pound compared to the cost of polyethylene at approximately thirty cents per pound. It is expected that in the future, increased consumer demand for biodegradable plastic products will result in increased use in consumer products, especially in specialty plastics applications.

One potential way to lower the cost of producing polyhydroxyalkanoate specialty products would be to lower raw materials costs. We posed the question of whether using organic waste as the carbon feedstock for bacterial polymer synthesis might be economically competitive with, or less costly than, using conventional substrates such as glucose or propionic acid, since the waste carbon source would be nearly cost-free.

The economic analysis examined the cost of using a waste acetic acid stream as the feed to the PHB/V fermentation process, and compared it to the cost of using glucose as a conventional substrate for PHB/V synthesis (*14*). In order to obtain copolymers containing valerate units, carbon chains of at least C3 must be supplied as substrate. In this example, propionic acid was also added to the process to yield PHB/V copolymer. The waste stream modeled in this example consisted of approximately 1.2% acetic acid and 96% water, the balance of the stream consisting of less than 1% each of several monomers used in acrylate manufacture. The yearly volume of the stream was calculated to generate a theoretical yield of 8.5 million

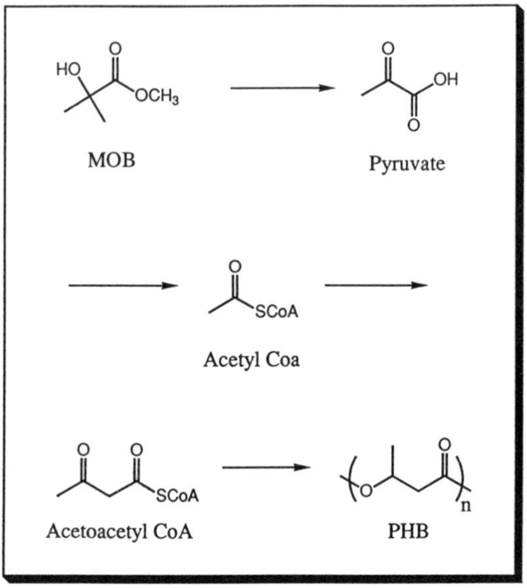

Overall Conversion

2 MOB ——————▶ 1 Acetoacetyl CoA

Figure 3. Hypothesized metabolic pathway for incorporation of methyl 2-hydroxyisobutyric acid to polyhydroxybutyrate polymer.

pounds PHB/V per year, based upon 34% conversion of waste organic carbon to polymer.

The manufacturing process used in the model was a two-stage aerobic fermentation system as described by Johnson (*14*). In the example provided in Johnson, the cost of PHB/V production from glucose and propionic acid was calculated to be approximately $2.32 per pound, which would include cost of investment in a fermentation facility, raw materials, and labor. When waste organic is used as substrate, the cost per pound of manufacture of PHB/V was calculated to be approximately $ 1.93 per pound. Factors in the calculations included raw materials costs (fermentation ingredients and solvents for polymer extraction), processing costs (utilities and labor) and fixed costs including depreciation, maintenance, and property overhead.

Discussion

Billions of pounds of organic carbon are discharged each year by chemical and agricultural manufacturing plants to waste treatment facilities to be degraded by biological, chemical or thermal processes. Discharges that might be amenable to treatment using bioconversion include waste streams generated by the chemical manufacture of polymers and plastics, or wastes produced in processing of dairy and vegetable food products. This study examined the feasibility of treating the organic carbon in waste streams by fermentation in order to convert waste carbon to either useful raw materials for the manufacture of biodegradable plastics, or to nonhazardous biodegradable solids that could be safely disposed of in landfills or used in land applications as a soil amendment. Bacteria obtained from the sludge generated by the trickle filter waste treatment system at an acrylate manufacturing site demonstrated the considerable advantage of selecting organisms already adapted to survival and vigorous growth in the presence of organic substrates. In addition, a number of bacterial strains obtained from the sludge were capable of synthesizing PHB from the waste carbon when nitrogen in the nutrient medium was limiting.

The technical and economic feasibility of a full-scale bioconversion system as proposed here that would could be used to treat industrial waste relies on addressing two fundamental concepts. First, bacterial strains that could produce the desired polymers under harsh environmental conditions must be developed, and second, a physical plant and engineering process must be designed, delivering waste at appropriate rates, permitting adequate rates of bacterial polymer synthesis, and providing processes for harvesting polymer and disposing of fermentation waste products.

A highly efficient polymer production system would most likely be required to process industrial waste streams which are typically produced in large volumes on a continuous or semi-continuous basis. A two-stage fermentation process, such as the one used in this study, would likely not be a suitable production system. However, constitutive PHA synthesis under non nutrient-limiting conditions has been demonstrated (*16,17*), and provides the potential for an economical option for waste bioconversion. As more is learned about the fundamental genetic mechanisms controlling both the rate of polymer synthesis and the mechanisms for substrate utilization that determine polymer composition, the potential to create genetically modified bacterial strains capable of high rates of specialty polymer production will also increase.

There is potential as well to produce polymers having unusual chemical properties by altering the composition of the feedstocks supplied to the conversion process. Bacterial polymer composition may be determined by substrate composition in the fermentation media, and copolymers of predictable composition can be produced by varying the both proportions and types of carbon substrates supplied to the process (5). The potential to vary polymer composition by supplementing waste streams with other carbon feedstocks affords the opportunity to manufacture specialty copolymers with varied thermoplastic properties.

Acknowledgments

The authors thank Yi Hyon Paik for preparing the chemical structures used in the figures.

Literature cited

1) Schlegel, H.G., A. Steinbuchel, Eds. *International Symposium on Bacterial Polyhydroxyalkanoates.* FEMS Microbiology Reviews 103; Göltze-Drück, Göttingen, Germany, **1992**.

2) Steinbuchel, A. In: *Biomaterials,* D. Byrom, MacMillan Publishers, Ltd. Great Britain, **1991**, 125-208.

3) Steinbuchel, A. *Acta. Biotechnol.* **1991**, 5,419-427.

4) Brandl, H, R. A. Gross, R.W Lenz, R.C. Fuller. *Adv. Biochem. Eng./Biotechnol.,* **1990**, 41, 77-93.

5) Doi, Y. *Microbial Polyesters* VCH Publishers, Inc. New York **1989**.

6) Kim, Y.B., R.W. Lenz, R.C. Fuller. *Macromol.* **1992**, 54, 1852-1857.

7) Fritsche K., R. Lenz and C. Fuller. *Int. J. Biol. Macromol.* **1990**, 12, 92-101 .

8) Fritsche K., R. Lenz and C. Fuller. *Makromol. Chem.* **1990**, 191, 1957-1965.

9) Doi, Y., Abe, C. *Macromol.* **1990**, 23, 3705-3707.

10) Lageveen, R.G., G.W. Huisman, H. Preusting, P. Kekelaar, G. Eggink, B. Witholt. *Appl. Env. Microbiol.* **1988**, 38, 1-5.

11) Brandl, H., E.J. Knee, R.C. Fuller, R.A. Gross, R.W. Lenz. *Int. J. Biol. Macromol.* **1989**, 11, 49-55.

12) Gross, R.A., C. Demello, R.W. Lenz, H. Brandl, R.C. Fuller. *Macromol.* **1989**, 22, 1106-1115.

13) Brandl G, Gross, R.A., Lenz, R.W., Fuller, R.C.. *Appl. Env. Microbiol.* **1988**, 54, 1977-1982.

14) Johnson H. E.   *Environmentally Degradable Polymers.*   Report No. 115A. Stamford Reseach Institute. April **1991**.

15) Wallen, L.L. and W.K. Rohwedder. *Env. Sci. and Technol.* **1974**, 8, 576-579.

16) Huisman G.W., E. Wonink, G. DeKoning, H. Preusting, B. Witholt . *Appl. Microbial. Biotechnol.* **1992**, 38, 1-5.

17) Page W.J., Knosp, O. *Appl. Env. Microbiol.* **1989**, 55, 1334-1339.

RECEIVED July 7, 1994

Chapter 15

Application of Lesquerella Oil in Industrial Coatings

S. F. Thames, M. O. Bautista, M. D. Watson, and M. D. Wang

Department of Polymer Science, University of Southern Mississippi, Box 10076, Hattiesburg, MS 39406-0076

Lesquerella oil (LO), a C_{20} triglyceride homologue of castor oil, is obtained from the seeds of *Lesquerella fendleri*. LO is being evaluated as a potential castor oil substitute. Novel polyesters containing LO have been synthesized and evaluated. Moreover, the functional LO polyesters have been used in the preparation of polyester-polyurethanes which have subsequently been evaluated for their performance as coatings compositions of high quality.

The genus, *Lesquerella*, is a "new" domestic crop with potential industrial significance. Twenty-three species of lesquerella have been evaluated, and of this number, *L. fendleri* has the most attractive agronomic potential (*1*). *L. fendleri*, a winter annual and native to the southwest, has seed containing over 25% oil by weight. Approximately 55% of that amount is lesquerolic acid, a C_{20} hydroxy fatty acid (*2,3*), while 20 to 36% is oleic, linoleic and linolenic acids in roughly a 2:1:2 ratio.

Lesquerella is particularly attractive as a raw material in that it, unlike other domestic vegetable oils, is structurally similar to castor oil because of its hydroxyl functionality (*4-7*). Hydroxyl containing fatty acids are unique and engender special rheological and reactivity properties. At present, castor oil and its derivatives are the sole commercial source of hydroxyl fatty acids (*8-10*). However, since the early 1970's, the castor oil used in this country has been imported. Annual U.S. imports have ranged from 29,000 to 64,000 metric tons (*11*). Price instability and inconsistent supply have handicapped end-users, thus making corporate planning difficult. Companies have sought alternative materials which include petroleum-based feedstocks. However, concerns for diminishing petroleum reserves, trade imbalances, the environment, and the needs of rural economies highlight the importance of developing and commercializing alternate sources of hydroxy fatty acids. Thus, a

0097–6156/94/0575–0212$08.00/0

domestic, economically-attractive and high-performance castor oil "substitute" is desirable, and it is for this reason that we are investigating the properties of LO.

Three hydroxy fatty acids (Figure 1), lesquerolic, densipolic, and auricolic acids, have been identified as the primary fatty acids in the seed oils of lesquerella species (*12-16*). It is normal for one acid to predominate over the other two within a species; however, all are similar to ricinoleic acid (C_{18}), the primary fatty acid of castor oil (*17*), yet they are C_{20} fatty acids and will reflect these structural differences in their properties.

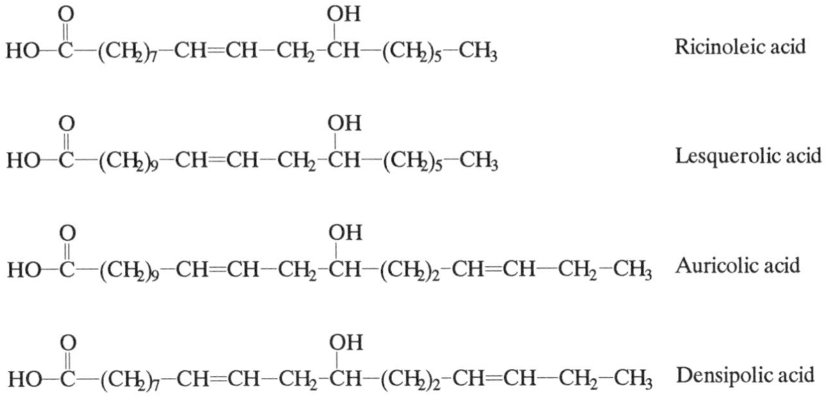

$$HO-\underset{\underset{O}{\|}}{C}-(CH_2)_7-CH=CH-CH_2-\underset{\overset{OH}{|}}{CH}-(CH_2)_5-CH_3 \qquad \text{Ricinoleic acid}$$

$$HO-\underset{\underset{O}{\|}}{C}-(CH_2)_9-CH=CH-CH_2-\underset{\overset{OH}{|}}{CH}-(CH_2)_5-CH_3 \qquad \text{Lesquerolic acid}$$

$$HO-\underset{\underset{O}{\|}}{C}-(CH_2)_9-CH=CH-CH_2-\underset{\overset{OH}{|}}{CH}-(CH_2)_2-CH=CH-CH_2-CH_3 \qquad \text{Auricolic acid}$$

$$HO-\underset{\underset{O}{\|}}{C}-(CH_2)_7-CH=CH-CH_2-\underset{\overset{OH}{|}}{CH}-(CH_2)_2-CH=CH-CH_2-CH_3 \qquad \text{Densipolic acid}$$

Figure 1. Chemical structures of hydroxyl containing fatty acids.

This work reports the utility of Lesquerella oil (LO) from *L. fendleri* as a raw material for industrial uses, particularly in flexible paints and coating systems. The LO's C_{20} fatty acid, lesquerolic acid, was expected to contribute desirable characteristics of hydrophobicity and flexibility to film-forming compositions. We have prepared polyesters, alkyd resins, containing LO as a co-reactant with a variety of reactive functional groups. Thus, a variety of coatings types were prepared, cured and tested as polyester polyurethanes.

Experimental Section

Synthesis of Alkyd Resins Based on LO. Polyesters of varying oil content, i.e. long and short oil alkyds were produced from lesquerella oil (LO), trimethylol ethane (TME), and phthalic anhydride(PA) (Table I). They were prepared by a two-step process involving glycerolysis and subsequently esterification (Scheme I).

Scheme I

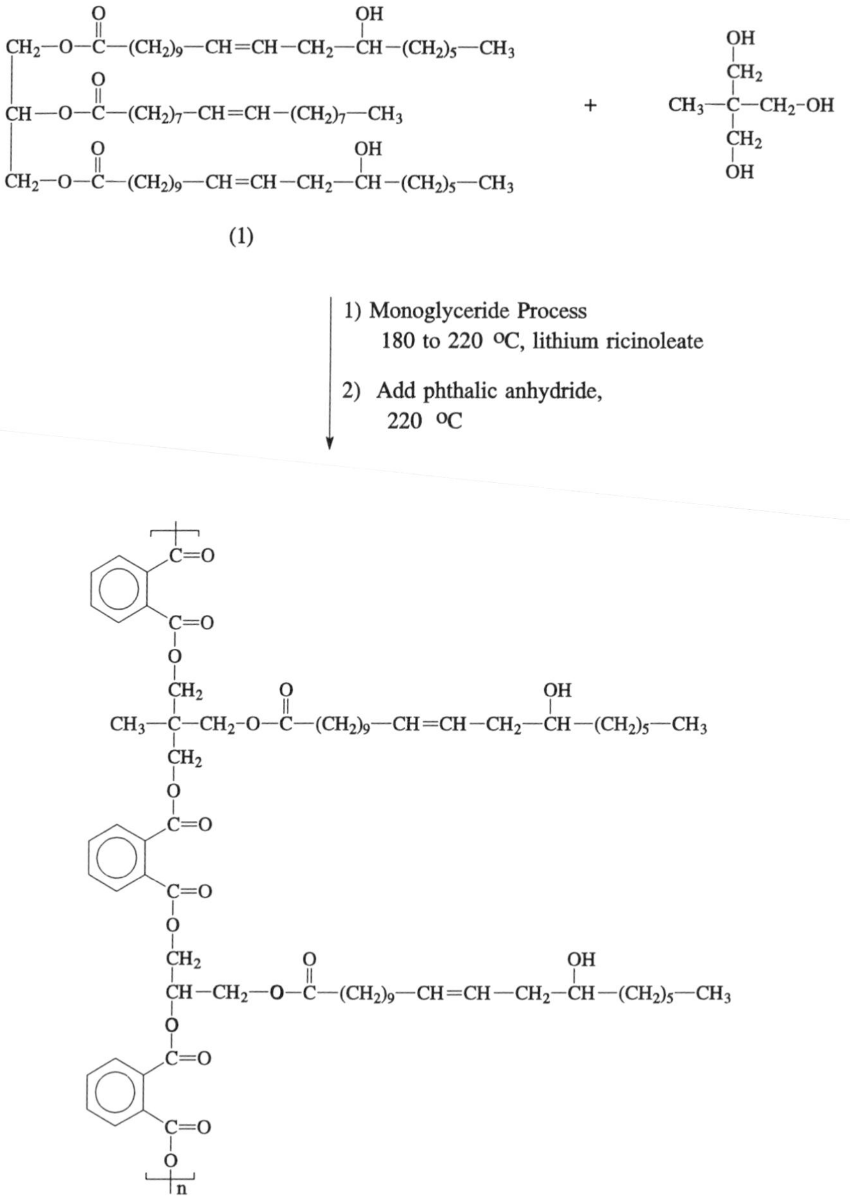

$$CH_2-O-\overset{\overset{O}{\|}}{C}-(CH_2)_9-CH=CH-CH_2-\overset{\overset{OH}{|}}{CH}-(CH_2)_5-CH_3$$

$$CH-O-\overset{\overset{O}{\|}}{C}-(CH_2)_7-CH=CH-(CH_2)_7-CH_3$$

$$CH_2-O-\overset{\overset{O}{\|}}{C}-(CH_2)_9-CH=CH-CH_2-\overset{\overset{OH}{|}}{CH}-(CH_2)_5-CH_3$$

$+$

$$CH_3-\overset{\overset{OH}{|}}{\underset{\underset{OH}{|}}{\overset{\overset{CH_2}{|}}{C}-CH_2-OH}}$$
$$\underset{CH_2}{}$$

(1)

1) Monoglyceride Process
   180 to 220 °C, lithium ricinoleate

2) Add phthalic anhydride,
   220 °C

$$CH_3-\overset{}{\underset{}{C}}-CH_2-O-\overset{\overset{O}{\|}}{C}-(CH_2)_9-CH=CH-CH_2-\overset{\overset{OH}{|}}{CH}-(CH_2)_5-CH_3$$

$$CH-CH_2-O-\overset{\overset{O}{\|}}{C}-(CH_2)_9-CH=CH-CH_2-\overset{\overset{OH}{|}}{CH}-(CH_2)_5-CH_3$$

Alkyd Resin via Alcoholysis Process

(2)

Table I. Formulations and Properties of Alkyd Resins

Alkyd	A	B	C
Lesquerella Oil, g	80.0	100.0	100.0
Trimethylol Ethane, g	55.0	26.4	86.7
Phthalic Anhydride, g	65.0	24.0	80.0
Lithium Ricinoleate, g	0.04	0.05	0.05
Properties			
% Nonvolatile Materials	94.4	94.5	93.1
Acid Value	11	8	9
OH Value	140	168	228
% Oil Length	40	66	37

First, LO was dissolved in xylene and heated to 180 °C. Then, TME and the lithium ricinoleate catalyst were added. The temperature was increased and held at 220 °C until sufficient alcoholysis occurred, i.e., when one volume of the reaction mixture forms a clear solution with three volumes of methanol. The glycerolysis process typically required one hour. When glycerolysis was completed, the reaction mixture was cooled to 180 °C, and the reaction vessel was fitted with a Dean-Stark trap for water removal during the esterification step. PA was added, and the temperature was increased to 220 °C. Aliquots were removed regularly and titrated to a phenolphthalein end point with methanolic KOH to determine an acid value. In doing so, the extent of the reaction was followed and terminated at an acid value of 11 or below. The hydroxyl value was determined according to method II.D.14 of IUPAC Standard Methods for the Analysis of Oils, Fats, and Derivatives (*18*). The percent solids of each alkyd was determined simply by measuring the weight loss of aliquots after drying to a constant weight. The properties of each of the alkyd resins are shown in Table I.

Formulation of Polyurethanes. Four polyfunctional isocyanates (Table II) were used to prepare eleven polyester-polyurethanes (Table III-V). Metal driers were included in an effort to supplement the drying process through oxidative polymerization of linoleic and linolenic acids. The resins were prepared using a 10% excess of isocyanate groups to hydroxyls. In the preparation of a typical formulation, all of the ingredients excluding the isocyanate(s) were combined and mixed with a high speed shaker. One hour before use, the isocyanate was added, and the blend was again thoroughly mixed. Samples were prepared for testing via application as a 7-mil wet film to cleaned and etched aluminum Q panels. The panels were prepared via treatment with a chromic/sulfuric acid solution as described in ASTM

D1730-67 TYPE B METHOD 2. The films were air-dried for 15 min to allow solvent flash-off and were subsequently heated at 110°C for 10 min. The samples were allowed to equilibrate for seven days before testing began.

Table II. Description of the Different Isocyanates

Trade Name	Vendor	Eq. Wt.	% Solids	Description
Desmodur N-3390	Miles	215	90	Isocyanurate of HDI
Desmodur CB-75	Miles	323	75	Isocyanurate of TDI
Desmodur HL	Miles	400	60	Polyisocynaurate from HDI and TDI
IPDI T1890L	Huls	350	70	Isocyanurate of IPDI

Table III. Formulations of Polyurethane Coatings with Alkyd Resin A

Formulation #	1	2	3	4
Alkyd A, g	7.4977	7.0004	7.0081	6.75
Des N-3390, g	4.4363	2.5851	2.7141	0
Des CB-75, g	0	2.5863	0	5.985
Des HL, g	0	0	0	0
IPDI T1890L, g	0	0	2.7149	0
Metacure T-12, g	0.1466	0.1457	0.1465	0
6% Cobalt[a], g	0.0265	0.0239	0.0229	0.0212
12% Zirconium[a], g	0.0224	0.0206	0.0234	0.0199
Solvent[b], g	6.3414	5.9286	6.5677	5.3312
BYK 306, g	0.0947	0.0937	0.0918	0.0905
Silwer 7602, g	0.0947	0.0909	0.0932	0.0905
% NVM	60	60	60	60

[a]0.02% Co and 0.0375% Zr (g/g Alkyd).
[b]10% n-Butyl acetate, 10% ethyl acetate, 25% methyl isobutyl ketone, 55% methyl ethyl ketone.

Table IV. Formulations of Polyurethane Coatings with Alkyd Resin B

Formulation #	5	6	7	8
Alkyd B, g	7.5011	6.7532	6.7523	6.7522
Des N-3390, g	5.3141	0	0	0
Des CB-75, g	0	7.1802	0	0
Des HL, g	0	0	8.894	0
IPDI T1890L, g	0	0	0	7.7852
Metacure T-12, g	0.1588	0.0825	0	0.1575
6% Cobalt[a], g	0.0243	0.0226	0.0208	0.0227
12% Zirconium[a], g	0.0215	0.0196	0.0215	0.0206
Solvent[b], g	5.2528	3.9893	2.3414	3.6232
BYK 306, g	0.0915	0.0915	0.0937	0.0914
Silwer 7602, g	0.0915	0.0909	0.0918	0.0914
% NVM	65	65	65	65

[a]0.02% Co and 0.0375% Zr (g/g Alkyd).
[b]10% n-Butyl acetate, 10% ethyl acetate, 25% methyl isobutyl ketone, 55% methyl ethyl ketone.

Table V. Formulations of Polyurethane Coatings with Alkyd Resin C

Formulation #	9	10	11
Alkyd C, g	6.002	5.5002	5.5015
Des N-3390, g	5.7735	0	0
Des CB-75, g	0	0	0
Des HL, g	0	9.8351	0
IPDI T1890L, g	0	0	8.6508
Metacure T-12, g	0.16	0	0.0713
6% Cobalt[a], g	0.02	0.0167	0.0175
12% Zirconium[a], g	0.0186	0.02	0.0175
Solvent[b], g	6.1547	3.9769	4.4477
BYK 306, g	0.0909	0.0987	0.0925
Silwer 7602, g	0.0885	0.0975	0.0982
% NVM	60	57	60

[a]0.02% Co and 0.0375% Zr (g/g Alkyd).
[b]10% n-Butyl acetate, 10% ethyl acetate, 25% methyl isobutyl ketone, 55% methyl ethyl ketone.

Results and Discussion

Formulations 4, 7, and 10 were excluded from testing as the former two were brittle to the point of flaking from the substrate with light fingernail pressure. Formulation 10 was clear when diluted to 57% nonvolatile material (NVM), but it produced opaque, cream colored dry films. This was likely the result of incompatibility between the isocyanate and the short oil alkyd. Formulations 2 and 3 utilized mixtures of isocyanates in an effort to obtain hardness with acceptable flexibility. The gel time (pot life), taken as the time for a complete formulation to gel in a closed container, was found to be greater than 8 h in all 11 formulations.

The test results are shown in Tables VI and VII. The scratch hardness for coatings formulated from the long oil alkyd (alkyd B, Table IV) were understandably lower than those for the shorter oil alkyds, alkyds A (Table III) and C (Table V). Generally, long oil alkyds combined with aliphatic isocyanates give flexible and softer films than those from shorter oil alkyds. And, these results are no exception.

While polyesters A and C have almost identical oil lengths, polyester C is higher in hydroxyl value. It therefore has the potential to affect higher crosslink densities when combined with isocyanate(s). For example, coating 9 gave 350 methyl ethyl ketone (MEK) double rubs compared to coating 1 with 150 MEK double rubs.

Table VI. Properties of Polyurethane Coatings

Formulation #	1	2	3	5	6	8	9	11
Conical Mandrel[a] (1/8") (ASTM D-522)	Pass	Pass	Pass	Pass	Pass	Pass	Pass	Fail
Crosshatch Adhesion[a] (ASTM D-3359)	5B	5B	5B	5B	5B	5B	5B	5B
Film Gouge Hardness by Pencil[a] (ASTM D-3363)	9H	9H	9H	7H	9H	9H	9H	8H
Film Scratch Hardness by Pencil[a] (ASTM D-3363)	HB	HB	HB	B	B	B	HB	HB
Specular Gloss[b] (ASTM D-523)	86 (20°) 99 (60°) 99 (85°)	94 (20°) 100(60°) 100(85°)	88 (20°) 100(60°) 99 (85°)	85 (20°) 100(60°) 101(85°)	97 (20°) 107(60°) 96(85°)	84 (20°) 99 (60°) 100(85°)	87(20°) 102(60°) 101(85°)	93 (20°) 106(60°) 101(85°)

[a] Average value from three samples.
[b] Average value from five samples.

Table VII. Chemical and Solvent Resistance of Polyurethane Coatings[a]

Formulation #	1	2	3	5	6	8	9	11
MEK Double Rubs (ASTM D-4752)	150 (2 mil)	230 (2 mil)	90 (2mil)	100 (2mil)	90 (2.4mil)	70 (2.6 mil)	350 (2.3mil)	50 (2.4 mil)
1 hour SpotTest,[b] Covered (ASTM D-1308)								
D.I. water	5	5	5	5	5	5	5	5
50% Ethyl Alcohol	5	5	5	5	5	5	5	5
Vinegar	5	5	5	5	5	5	5	5
Conc. NH_4OH	5	5	5	5	5	5	5	5
20% H_2SO_4	5	5	5	5	5	5	5	5
Soap Solution	5	5	5	5	5	5	5	5
Mineral Spirits	5	5	5	5	5	5	5	5
Coffee	5	5	5	5	5	5	5	5
Corn Oil	5	5	5	5	5	5	5	5

[a]Average value from three samples.
[b]5 = no effect, 4 = stains only, 3 = blistering, 2 = lifted film, 1 = failure

When the short oil alkyd A was used to prepare Formulation 2 with a mixture of Desmodur N-3390 and CB-75, improved performance in MEK double rubs and gloss was noted over Formulation 1, which used only N-3390. However, the opposite was true for Formulation 3 which used a mixture of N-3390 and IPDI T1890L.

The data of Figure 2 confirm the structure-T_g relationship of higher T_g's for urethane-short oil alkyds, i.e. the increases in T_g accompany decreases in oil length. Polyurethane-polyesters of aromatic isocyanate content (Formulations 2 and 6) are higher in T_g than coatings derived from aliphatic isocyanates, while the reverse holds true for thermal degradation (Formulations 2 and 6).

All films showed excellent gloss and chemical resistance and, with the exception of Formulation 11, all were high in flexibility.

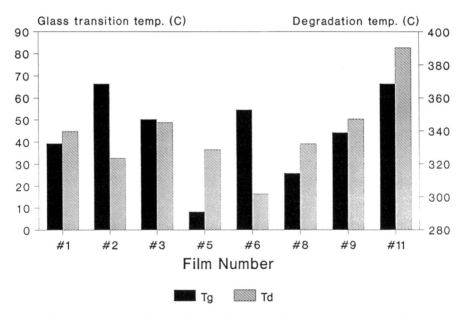

Figure 2. Glass transition (T_g) and degradation temperatures (T_d) of polyurethane alkyd coatings.

Conclusion

Polyurethane-polyesters of lesquerella oil content were synthesized, formulated, cured and tested for physical and chemical properties. When the formulations balanced oil length, hydroxyl value, isocyanate type and equivalence, superior properties were obtained. These included hardness, specular gloss, chemical and solvent resistance, as well as superior flexibility and adhesion. The coatings are attractive as clear finishes, and from a formulation stand point, their 60%+ non-volatiles, i.e. low volatile organic compounds (VOC's), is attractive.

Acknowledgments

We thank Mr. James Brown, President of International Flora Technologies, LTD., for supplying the lesquerella oil. This research is based upon work supported by the Cooperative State Research Service, U.S. Department of Agriculture, under Agreement No. 91-38202-5928 administered by Dr. Daniel E. Kugler, Dr. Davis Clements and Mrs. Carmela Bailey.

Literature Cited

1. Roetheli, J. C.; Carlson, K. D.; Kleiman, R.; Thompson, A. E.; Dierig, D. A.; Glaser, L. K.; Blase, M. G.; Goodell, J. **Lesquerella as a Source of Hydroxy Fatty Acids for industrial Products**, USDA, Cooperative State Research Service, October 1991.
2. Carlson, K. D.; Chaudhry, A.; Bagby, M. O. *J. Am. Oil Chem. Soc.*, **1990**, *67*, p 438.
3. Carlson, K. D.; Chaudhry, A.; Peterson, R. E.; Bagby, M. O. *J. Am. Oil Chem. Soc.*, **1990**, *67*, p 495.
4. Barclay, A. S.; Gentry, H. S.; Jones, Q. *Economic Botany*, **1962**, *16*, p 95.
5. Gentry, H. S.; Barclay, A. S. *Economic Botany*, **1962**, *16*, p 206.
6. Thompson, A. E.; Dierig, D. A. *El Guayulero*, **1988**, *10*, p 16.
7. Thompson, A. E. In **Arid Lands: Today and Tomorrow,** Whitehead, E. E. et al, Ed.; Westview Press: Boulder, CO, 1988; pp 1311-1320.
8. Technical Bulletin No. 2, **The Chemistry of Castor Oil and Its Derivatives and Their Applications**, International Castor Oil Association, Westfield, NJ.
9. Achaya, K. T. *J. Am. Oil Chem. Soc.*, **1971**, *48*, p 758.
10. Naughton, F. C. *J. Am. Oil Chem. Soc.*, **1973**, *51*, p 65.
11. Roetheli, J. C.; Glaser, L. K.; Brigham, R. D. **Castor: Assessing the Feasibility of U.S. Production**, USDA, Cooperative State Research Service, April 1991.
12. Smith Jr., C. R.; Wilson, T. L.; Miwa, T. K.; Zobel, H.; Lohmar, R. L.; Wolff, I. A. *J. Org. Chem.*, **1961**, *26*, p 2903.
13. Mikolajczak, K. L.; Earle, F. R.; Wolff, I. A. *J. Am. Oil Chem. Soc.*, **1962**, *39*, p 78.
14. Smith Jr., C. R.; Wilson, T. L.; Bates, R. B.; Scholfield, C. R. *J. Org. Chem.*, **1962**, *27*, p 3112.
15. Miller, R. W.; Earle, F. R.; Wolff, I. A.; Jones, Q. *J. Am. Oil Chem. Soc.*, **1965**, *42*, p 817.
16. Kleiman, R.; Spencer, G. F.; Earle, F. R.; Nieschlag, H. J.; Barclay, A. S. *Lipids*, **1972**, *7*, p 660.
17. Smith, Jr., C. R. In *Fatty Acids*, Pryde, E. H., Ed.; American Oil Chemists' Society: Champaign, IL, 1979, pp 29-47.
18. Paquot, C. In *IUPAC Standard Methods for the Analysis of Oils, Fats and Derivatives*; 6th edition; Pergamon Press: Oxford, UK, 1979; pp 89-91.

RECEIVED May 24, 1994

Chapter 16

Synthesis, Characterization, Derivation, and Application of Guayule Coproducts

A Review

S. F. Thames, P. W. Poole, Z. A. He, and J. K. Copeland

Department of Polymer Science, University of Southern Mississippi,
Box 10076, Hattiesburg, MS 39406-0076

Guayule (*Parthenium Argentatum* Gray), a shrub growing in the southwestern United States and northern Mexico, is a promising native source of natural rubber. Processing of the guayule shrub provides five coproduct fractions including high molecular weight rubber, low molecular weight rubber, organic soluble resin, water soluble resin, and bagasse. The work herein will focus on the synthesis and evaluation of low molecular weight guayule rubber (LMWGR) derivatives such as chlorinated hydroxylated LMWGR, acrylated chlorinated LMWGR, epoxidized LMWGR, and maleinized LMWGR. In particular, high solids coatings, 100% solids UV cured coatings, and water-borne coating applications will be discussed.

A domestic supply of natural rubber is important to a modern industrialized economy. Among rubber producing plants, guayule provides high quality, high molecular weight natural rubber with properties comparable to *Hevea Brasiliensis* (Malaysian rubber)(*1*). Due to cultivation, harvesting, and processing costs, the value and quantity of high molecular weight guayule rubber is insufficient to justify production. Therefore, the development of value-added materials from the coproduct fractions of guayule is necessary. We have shown that low molecular weight guayule rubber (LMWGR) can be a valuable precursor in high solids, UV cured, and water-borne coating formulations.

Coatings formulated with chlorinated rubber are well known for their excellent chemical and water resistance, good abrasion resistance, and good flame retardancy (*2*). As such, chlorinated rubber is used extensively in harsh environments for marine applications, swimming pool paints, and traffic paints (*2*). However, the solubility of chlorinated rubber has limited its use to solvent soluble coatings of high volatile organic content (VOC). Thus, technologies offering reductions in VOC are important and constitute value added products. Accordingly, we have developed novel 100% solids coatings that cure rapidly with UV light. They offer energy savings, ease of handling, and wide formulation latitude.

The addition of reactive groups to the polymer backbone of chlorinated rubber results in the formation of a convertible or reactive material that is useful in the preparation of environmentally compliant chlorinated rubber coatings. Thus, chlorinated hydroxylated LMWGR (ChLMWGR) has been synthesized, characterized, and utilized in high solids coatings applications. Higher solids polyurethane coatings thus formulated show properties superior to commercial chlorinated rubber products. Specifically, the crosslinking of ChLMWGR with a polyisocyanate based on hexamethylene diisocyanate (HDI) improves the solvent resistance properties of chlorinated rubber coatings.

Additionally, ChLMWGR has been reacted with acryloyl chloride to produce acrylated chlorinated rubber (ACR), a binder formulated into 100% solids UV cured coatings for use in wood fillers and clear finishes. They are characterized by excellent water resistance, good chemical resistance, and superior solvent resistance. Typical applications include use as wood coatings for gym floors, roller rink floors, and furniture finishes.

Epoxy resins have wide spread use in the coating industry (3). However, little attention has been given to the development of coating systems based on low molecular weight epoxidized rubber. Epoxidized LMWGR (EGR) is synthesized via the reaction of LMWGR and m-chloroperbenzoic acid. Epoxide content is variable in that it can be incorporated over a range of stoichiometric ratios from 5 to 100%.

The reaction of LMWGR, maleic anhydride and the free radical initiator, benzoyl peroxide (BPO) gives maleinized LMWGR. The maleinized LMWGR has been further modified with chlorin1xand inMwstigated as a component of water-borne coatings. Water dispersibility has been achieved via formation of the carboxylic acid functionalities after ring opened is affected.

Experimental Section

The OSR of guayule was received from Texas A&M and the crude LMWGR was isolated from OSR via acetone extraction, dissolution of crude rubber in methylene chloride, and subsequent precipitation into 95% ethanol. Pure LMWGR was dried to constant weight in a Napco vacuum oven at ambient temperature under reduced pressure. Reagent grade methylene chloride, chloroform, ethanol, and methanol were obtained from Fisher Chemicals while reagent grade sodium hydroxide, sodium methoxide, m-chloroperbenzoic acid, trichloroacetic acid, benzoyl peroxide, anhydrous toluene, p-xylene, maleic anhydride, and acryloyl chloride were purchased from Aldrich Chemicals. Triethylamine was supplied by BASF. Reagent grade acetic acid was obtained from Baker.

FTIR data were obtained with a Nicolet Systems IR42 Infrared Spectrometer. NMR data were obtained on a Bruker 300 mHz NMR in solution state. Solid state NMR spectra were obtained on a Bruker 300 mHz solid state NMR. Differential Scanning Calorimetry was performed on a Mettler Instruments DSC 30. Elemental analysis was performed by MHW laboratories, Phoenix AZ. Melting points were obtained with a Mel Temp laboratory device.

Chlorinated Hydroxylated Guayule Rubber. To 350 ml of 5% rubber in anhydrous toluene in a 500 ml Erlenmeyer Flask, 56.8 g of trichloroacetic acid was added. The solution was stirred for 23 h at 0 °C under N_2 atmosphere and transferred to a 1000 ml Erlenmeyer flask containing 60.7 g of sodium methoxide and 100 ml of methanol. After 20 h under stir, acetic acid (34.7 ml) was added to neutralize the mixture, after which stirring was continued for 15 min. The flask contents were poured into 650 ml of methanol and the pH adjusted to 5-6 with excess acetic acid. The hydroxylated guayule rubber product was thus precipitated, washed with methanol and 50% (v/v) methanol in water, and dried to constant weight. It was made into a 5% CH_2Cl_2 solution in a three neck flask fitted with gas inlet tube, stopper, water condenser, and magnetic stirrer. The inlet tube was connected to a chlorine cylinder through a gas trap via Teflon tubing. The exit port was attached to two successive traps filled with concentrated sodium hydroxide solutions. The flask was immersed in an oil bath for temperature control and the system was purged with N_2. Nitrogen flow was terminated after 15 min after which chlorine was bubbled through the mixture under reflux until Fourier transform infrared spectra indicated no further decrease in double bond absorptions. The product was precipitated with 600 ml of methanol and dried to constant weight.

Acrylated Chlorinated Guayule Rubber. Acryloyl chloride (2.178 g) was added to a solution of 11.0 g of ChLMWGR and 150 ml of anhydrous toluene in a 500 ml three neck flask equipped with nitrogen inlet tube, water condenser, and pressure equalizing separatory funnel. The mixture was stirred at 70 °C under N_2. Triethylamine (2.45 g) in 50 ml of anhydrous toluene was added dropwise over 1 h and the reaction was allowed to proceed for 2 h. The mixture was then precipitated in 800 ml of methanol and dried to constant weight.

Epoxidized Guayule Rubber. A molar amount of pure LMWGR (5% in CH_2Cl_2) was added to a three neck flask equipped with a pressure equalizing separatory funnel and placed on a magnetic stirrer. The funnel was charged with m-chloroperbenzoic acid corresponding to the desired mole percent epoxidation. The flask, charged with the LMWGR solution, was immersed in an ice bath and stirred for 15 min followed by the dropwise addition of acid solution. After complete addition, the reaction was allowed to stir for 30 min, filtered with a 44 X 36 Gardco extra fine paint filter followed by filtration over Celite filter aid, and the product was dried to constant weight by rotary evaporation.

Maleinized Guayule Rubber. Maleinization was performed in a four neck flask charged with a stoichiometric amount (corresponding to the desired percent maleinization) of maleic anhydride in *p*-xylene. The reaction was heated to 118 ± 1 °C and a solution of LMWGR in *p*-xylene was added dropwise via a pressure equalizing separatory funnel. After mixing for 5 min, a mixture of initiator (benzoyl peroxide) in *p*-xylene was added dropwise over one hour and the reaction was continued for 1 h and cooled to room temperature. The maleinized guayule rubber was then precipitated in petroleum ether and dried to constant weight and characterized via FTIR.

Chlorinated Maleinized Guayule Rubber. Maleinized guayule rubber (1 g) was dissolved in methylene chloride, added to a three neck flask equipped with a condenser, gas dispersion tube, and thermometer, and reacted with chlorine. The reaction set up is equivalent to the method used for chlorinated hydroxylated guayule rubber listed above. After reaction, the product was precipitated in petroleum ether and dried to constant weight.

Results and Discussion

Synthesis of ChLMWGR and ACR. Scheme 1 describes the synthetic route to ChLMWGR (I) and derivative, ACR (II). FTIR and NMR support the formation of I and II (Figures 1 and 2). The FTIR spectrum corroborate chlorine addition via strong absorptions for I and II (Figure 1) spectra at 738 cm^{-1}. and hydroxyl formation on I via absorptions at ca. 3400-3600 cm^{-1}. The loss of hydroxyl absorptions at 3400-3600 cm^{-1} and the appearance of characteristic acrylate absorption (1719, 1635, 1403, and 1198 cm^{-1}) confirm complete acrylation and thus the formation of II.

Similarly, NMR spectra are consistent with the formation of both I and II (Figure 2). Broad spectroscopic patterns below 80 ppm are quite similar and show structural complexity for the products. The broad absorptions at 18 and 59 ppm are contributed to methyl carbons (-CH$_3$), methylene carbons (-CH$_2$-), and methyl carbons bonded to a single chlorine (-CH$_2$Cl) (4). At 66 ppm the broad absorption is indicative of methylene carbons bonded to a single chlorine (-CHCl-). Additionally absorption at 75 ppm is representative of tertiary carbons bonded to chlorine (=C-Cl) (4). For II, deviations from I occur above 80 ppm. The absorptions at 84 ppm coincide with the tertiary carbon bonded to the acryloyl group. The broad absorption at ca. 131 ppm is attributed to the acryloyl vinyl carbons and the remaining vinyl units of the natural rubber backbone. Weaker absorptions at 165 ppm are characteristic of acryloyl carbonyl carbons. Despite extensive NMR data, the structure of I and II is unknown, a result of the complexity of the chlorinated structures. Tables I and II include chlorine and acrylate content, and glass transition temperatures.

Table I. Properties of Chlorinated Hydroxylated LMWGR

Hydroxyl content (Wt %)	1.8
Chlorine content (Wt %)	48.3

Table II. Properties of ACLMWGR

Chlorine content (Wt %)	46.4
Acryl content (Wt %)	7.0
Tg, °C	77.9

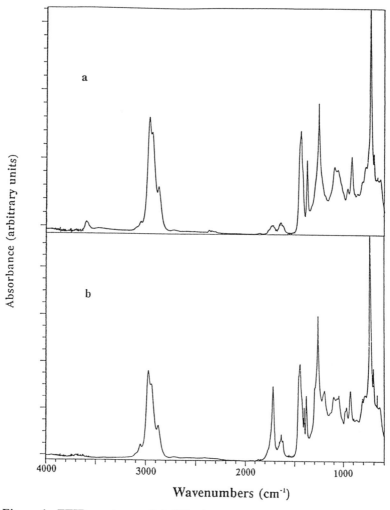

Scheme 1. Hydroxylation and Acrylation of Low Molecular Weight Guayule Rubber. (I) Chlorinated Hydroxylated LMWGR (II) Acrylated Chlorinated LMWGR.

Figure 1. FTIR spectrum of a) Chlorinated Hydroxylated LMWGR and b) Acrylated Chlorinated LMWGR.

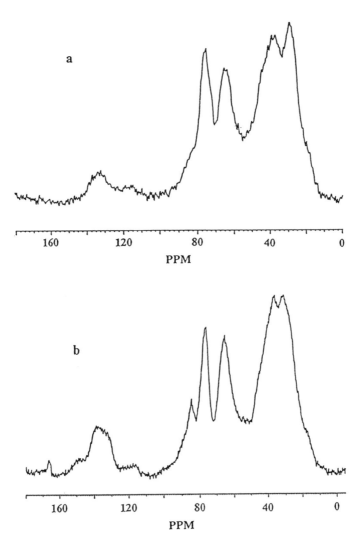

Figure 2. ^{13}C **Solid State NMR spectrum of a) Chlorinated Hydroxylated LMWGR and b) Acrylated Chlorinated LMWGR.**

To meet the requirements of a variety of end uses, I, liquid polyol and polyisocyanate were used as reactive diluents and thus increased the solids content of the coatings. The I (2.7 weight % hydroxyl), with polyol (Tone   301), was reacted with isocyanate (Desmodur   N-100) in a hydroxyl/isocyanate ratio of 1.0/1.05-1.10. The coating was cured at room temperature. Formulation and coating performance are noted in Table III and confirm toughness, high gloss, and resistance to water, organic solvents, and chemicals.

**Table III. Formulation of Clear Two Component
Polyurethane Coatings with ChLMWGR.**

Material	Amount (g)
ChLMWGR	39.80
Tone 301	9.70
Ektapro EEP	20.00
Xylene	25.00
Metacure T-120	0.15
Byk 080	0.20
Byk 325	0.20
Desmodur N-100	34.90

Property	Value
Wet thickness	2 mils
Drying time:	
Set to touch	15 min
Dust free	30 min
Tack free	105 min
Solid content	65.40
Pencil hardness	H
Tensile strength, Kpsi	3.90
Elongation at break	12%
Impact (in-lb) direct	120
and reverse	80
Adhesion (ASTM D-3359)	5B
MEK (double rub)	200
8 hour spot tests:	
Water	5
Concentrated NH4OH	5
10% NaOH	5
20% H2SO4	5

5 = no effect   4 = stain only   3 = blistering   2 = lifted film   1 = failure

II was used in both filler coatings and clear finishes for wood substrates. The combination of photoinitiators of cleavage type (Irgacure 651) and abstraction type (benzophenone) were used to enhance the curing efficiency. The hardness and flexibility of the coatings can be adjusted by careful selection of the reactive diluents. Increasing amounts of trifunctional diluents (e.g., Photomer 4094 and Photomer 4149 from Henkel) contributes high crosslinking density and leads to increased hardness with less flexibility. Due to the slight yellow color due to impurities in LMWGR, no stain is needed to achieve a rich wood-like tone. The fillers (Table IV) show excellent adhesion, good chemical, water, solvent resistance and excellent sanding properties. Ultraviolet cured finishes (Table V) are attractive, of high gloss, and possess excellent adhesion, good water and chemical resistance, and superior solvent resistance exhibited by 500 methyl ethyl ketone (MEK) double rubs.

Table IV. Matte UV Curing Acrylated Chlorinated ACR Wood Filler

Material	Amount (g)	Supplier
ACLMWGR	15.8	USM
Photomer 4061	50.0	Henkel
Photomer 4094	20.0	Henkel
Photomer 4770	5.0	Henkel
Byk 065	0.7	Byk
DisperByk 163	2.5	Byk
Microwhite 50	30.0	E.C.C.
Irgacure 651	1.0	Ciba-Geigy
Benzophenone	2.0	Dainippon

Ground to Hegman 7.5 with high speed mixer at 1500 rpm for 1 hr.

Property	Value	ASTM Method #
Viscosity @ 25 °C, cps	5,300	
Wet film thickness	2 mils	
Adhesion	5B	D-3359
Pencil Hardness	5H	D-3363
Tensile strength psi	4,500	D-2370
Elongation at break	8%	D-2370
MEK double rub	500+	D-4752
8 hour spot tests:		D-1308
Water	5	
Concentrated NH_4OH	4	
10% NaOH	4	
20% H_2SO4	5	

5 = no effect   4 = stain only   3 = blistering   2 = lifted film   1 = failure

Table V. High Gloss UV Curing ACLMWGR Wood Finishes

Material	Amount (g)	Supplier
AC **LMWGR**	22.0	USM
Photomer 4127	20.0	Henkel
Photomer 4061	20.0	Henkel
Photomer 4094	30.0	Henkel
Photomer 4149	6.0	Henkel
Photomer 4770	10.0	Henkel
Byk 065	0.7	Byk
Byk 325	1.5	Byk
Irgacure 651	2.0	Ciba-Geigy
Benzophenone	2.0	Dainippon

Property	Value	ASTM Method #
Viscosity @ 25 ^{0}C, cps	3,660	
Wet film thickness	1 mil	
Adhesion	5B	D-3359
Pencil hardness	4H	D-3363
Tensile strength psi	3,900	D-2370
Elongation at break	7%	D-2370
60^0 gloss	85	
MEK (double rub)	500+	D-4752
8 hour spot tests:		D-1308
Water	5	
Concentrated NH_4OH	4	
10% NaOH	4	
20% H_2SO_4	5	

5 = no effect   4 = stains only   3 = blistering   2 = lifted film   1 = failure

Epoxidized LMWGR (EGR). Scheme 2 represents the epoxidation of LMWGR. FTIR (Figure 3) data authenticates the formation of EGR due to the characteristic internal epoxide absorptions at 1255, 1067, and 874 cm^{-1} (5). Furthermore, the loss of absorbance observed at 3025 and 1660 cm^{-1} demonstrates loss of olefinic groups.

Proton (^{1}H) NMR (Figure 4) also confirms reaction and is used quantitatively to determine the percent epoxidation. The relationship between the area under the epoxy proton peak at 2.7 ppm (AE) and the olefinic proton peak at 5.1 ppm (AO) is given by the following equation in which the expected percent epoxidation is in all cases consistent with NMR data.

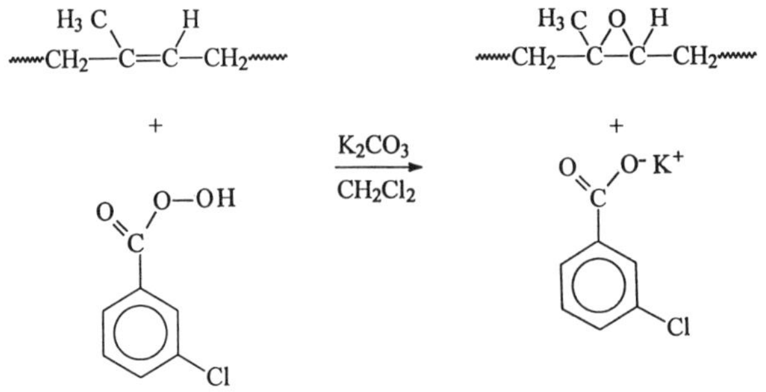

Scheme 2. Epoxidation Mechanism of LMWGR.

$$\%Epoxidation = \frac{\int AE}{\int AE + \int AO}$$

Polymer formation by ring opening reactions of epoxides are environmentally ideal as no VOC emissions result. Thus, EGR was evaluated as a reactant with carboxylic acids via isothermal differential scanning calorimetry (IDSC). The enthalpy of the reaction at various temperatures and reactant concentrations was measured to determine the reaction order and energy of activation. Accordingly, 20 mg samples of EGR and acid (0 to 10 mole % acid to oxirane content of EGR) were evaluated in the DSC at 60 °C. Samples were heated at 100 °C/min and held isothermally at a desired temperature for 60 minutes under nitrogen. All data curves were normalized to the same baseline for each acid type in order to compensate for differences between the sample and reference pan weights. The change in enthalpy with time is proportional to the change in the reactant concentration with time as the ring opening of epoxides with carboxylic acids gives chain modification via ester formation. A plot of the change in enthalpy with time versus acid concentration yields a line whose slope is equivalent to the reaction order. In all cases, the reaction order was determined as first order with respect to acid concentration. The heat flow for each sample at a given time is proportional to the change in concentration with time. In this instance, heat flow reached a maximum at approximately two minutes into the reaction, and a minimum after 30 to 60 minutes, when the concentration of free carboxylic acids is effectively zero. Since the maximum heat flow shown on the DSC thermogram does not correspond to the initial rate, the plot must be extrapolated to time zero to obtain the maximum initial heat flow and thus the initial rate (6). A plot of the natural log of the initial heat flow versus 1/T gives a line whose slope is equal to Ea/R, thus the energy of activation is determined (Figure 5).

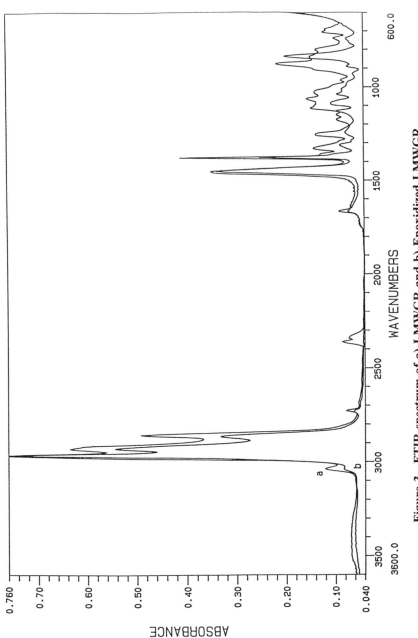

Figure 3. FTIR spectrum of a) LMWGR and b) Epoxidized LMWGR.

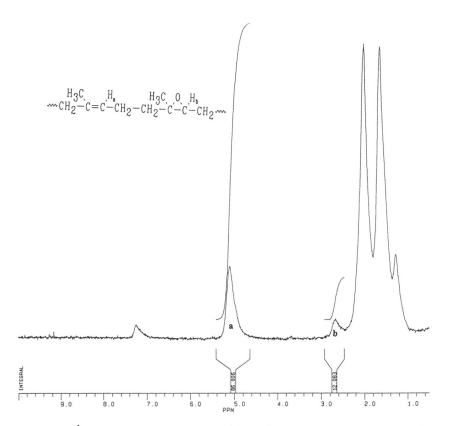

Figure 4. ^{1}H NMR spectrum of LMWGR a) vinylic proton absorbance and b) oxirane proton absorbance. (Reproduced with permission from ref. 7. Copyright 1993 John Wiley and Sons.)

The pKa values in water for stearic acid, meta-chlorobenzoic acid, and dichloroacetic acid are 4.95, 3.83, and 1.28, respectively. One might assume that the more acidic compound would be the most reactive and exhibit the lowest activation energy. However, this did not prove to be the case. An activation energy of 87.16 kJ/mol was determined for stearic acid and since its pKa is high, this low reactivity was expected. The activation energy of the meta-chlorobenzoic acid followed the expected trend with an activation energy of 42.90 kJ/mol, a considerably more reactive moiety than stearic acid. Dichloroacetic acid on the other hand, with an activation energy of 115.96 kJ/mol, was much less reactive than expected. Based on acidity alone, this acid was expected to have the lowest activation energy of the acids studied. One possible explanation of this unexpected observation is that the chlorine atoms hamper this sterically hindered reaction. Additionally, solubility and compatibility of the acid with the epoxidized rubber may hinder the reaction. Apparently the reactivity of the rubber is low; we suspect that this is due mainly to steric considerations and thus the choice of acid is critical. Therefore, the use of EGR as a crosslinking agent with carboxylic acids requires certain criteria to be successful. First, the temperatures to cure an epoxidized

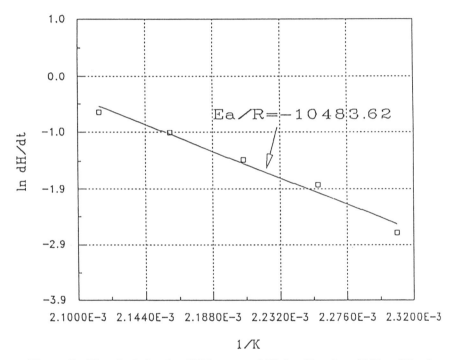

Figure 5. Plot depicting ln dH/dt versus 1/K for Stearic acid/Epoxidized LMWGR reactions. (Reproduced with permission from ref. 8. Copyright 1994 Federation of Societies for Coatings Technology.)

rubber system will be relatively high without the use of a catalyst. Secondly, carboxylic acid used successfully as crosslinking agents will require a pKa < 3.5 if a sufficient rate of cure is to be achieved. Finally, the reactive acid must be sterically unencumbered.

Maleinized Guayule Rubber (MGR). The IR spectra of MGR and chlorinated MGR are shown in Figures 6 and 7. These spectra confirm the presence of the MA graft site as strong absorption bands indicative of cyclic anhydride presence appear at 1857 cm^{-1} and 1780 cm^{-1}. For chlorinated MGR, a strong peak at 750 indicates the presence of the carbon-chlorine bonds. As the reaction proceeds, this peak reaches a maximum absorbance indicating reaction is complete. Peaks occurring at 2970, 1450, and 1217 cm^{-1} indicate the presence of carbon-hydrogen bonds in the natural rubber structure.

Solid state ^{13}C spectra of MGR and chlorinated MGR are summarized in Tables VI and VII. Resonance values agree with reported literature values (5). Carbon spectra are sensitive to substitution of hydrogen by chlorine. The resonances appeared broad due to a variety of possible chlorinated structures present as in chlorinated natural rubber.

Table VI: Resonance Assignments for MAGR

Resonances	Assignments
23.9, 34.7, 47.5	CH_2, CH_3
135.4, 130.1	$CH_2=CH_1$
172.9, 165.7	C=O

Efforts to produce a water borne coatings composition were successful. Accordingly, water (25g), butyl cellosolve (5g), chlorinated MGR (2.5g), amine crosslinking agents (1.75g), and EPON 828 (3.5g) formed continuous, hard, low VOC coating compositions possessing excellent water resistance properties. Films were cured at 150 °C overnight and placed in a water bath for six months with no change in coating properties.

Table VII: Resonance Assignments for CMGR

Resonances	Assignment	Literature
38.4, 45.6	CH_2Cl	45-48
63.3	CHCl	62-64
77.1	=C-Cl	74-77
90.3	$>CCl_2$	---

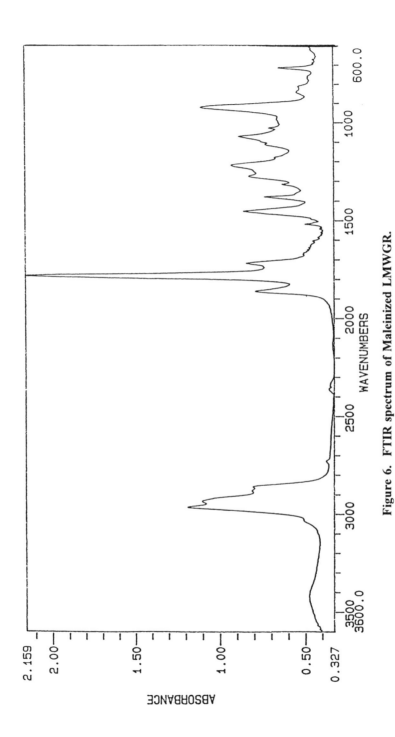

Figure 6.   FTIR spectrum of Maleinized LMWGR.

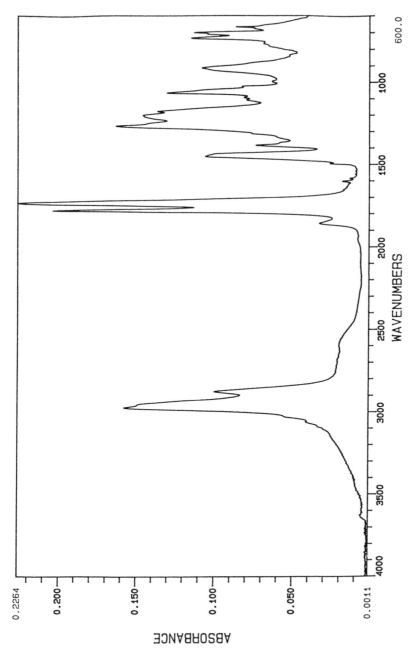

Figure 7. FTIR spectrum of Chlorinated Maleinized Guayule Rubber.

Conclusions

Guayule provides a promising source of natural rubber and as such is a valued commodity to the United States. Its coproduct fractions have been utilized in a variety of high performance, high quality coating applications including 100% solids UV cured, water-borne, higher solids, and solvent based coatings. The synthesis of low molecular weight guayule rubber derivatives include chlorinated hydroxylated, acrylated chlorinated, epoxidized, maleinized, and chlorinated maleinized low molecular weight guayule rubber has been accomplished. Their structures have been confirmed by via standard techniques.

Acknowledgments

The authors thank Dr. Skip McDaniel of the USDA Forestry Sciences Laboratory, Dr. William Walker and Mr. Steve Manning of the Gulf Coast Research Center, and Dr. Daniel E. Kugler and Mrs. Carmela Bailey of USDA/CSRS for their assistance and helpful advice. This research is based upon work supported by the Cooperative State Research Service, U.S. Department of Agriculture, under Agreement No. 92-COOP-1-6921.

Literature Cited

1.   *Guayule Natural Rubber*: Whitworth, J. W.; Whitehead, E. E., Eds.; USDA, Office of Arid Lands Studies, The University of Arizona: Tucson, AZ, 1991; pp 12-14.
2.   Paul, S. *Surface Coatings*, John Wiley and Sons: Chichester, NY, 1985; pp 248-260.
3.   The Oil and Colour Chemist's Association *Surface Coatings I: Raw Materials and Their Uses*; Chapman and Hall: London, 1983; pp 120-129.
4.   Makani, S., Brigodiot, M., and Marechal, E. *J. Appl. Poly. Sci.*, 29, **1984**, p 4081.
5.   Silverstein, R. M., Bassler, G. C., and Morrill T. C. *Spectrometric Identification of Organic Compounds*, John Wiley and Sons: New York, NY, 1981; pp 420-423.
6.   Gan, S. N. and Burfield, D. R. *Polymer*, 60, **1989**, p 1903.
7.   Thames, S. F. and Poole, P. W. *J. Appl. Poly. Sci.*, 47, **1993**, pp 1255-1262.
8.   Thames, S. F. and Copeland, J. K., *J. Coat. Tech.*, 66(833), **1994**, pp 59-62.

RECEIVED May 24, 1994

Author Index

Affiliation Index

Subject Index

Bestsellers from ACS Books

The ACS Style Guide: A Manual for Authors and Editors
Edited by Janet S. Dodd
264 pp; clothbound ISBN 0–8412–0917–0; paperback ISBN 0–8412–0943–X

The Basics of Technical Communicating
By B. Edward Cain
ACS Professional Reference Book; 198 pp;
clothbound ISBN 0–8412–1451–4; paperback ISBN 0–8412–1452–2

Chemical Activities (student and teacher editions)
By Christie L. Borgford and Lee R. Summerlin
330 pp; spiralbound ISBN 0–8412–1417–4; teacher ed. ISBN 0–8412–1416–6

Chemical Demonstrations: A Sourcebook for Teachers,
Volumes 1 and 2, Second Edition
Volume 1 by Lee R. Summerlin and James L. Ealy, Jr.;
Vol. 1, 198 pp; spiralbound ISBN 0–8412–1481–6;
Volume 2 by Lee R. Summerlin, Christie L. Borgford, and Julie B. Ealy
Vol. 2, 234 pp; spiralbound ISBN 0–8412–1535–9

Chemistry and Crime: From Sherlock Holmes to Today's Courtroom
Edited by Samuel M. Gerber
135 pp; clothbound ISBN 0–8412–0784–4; paperback ISBN 0–8412–0785–2

Writing the Laboratory Notebook
By Howard M. Kanare
145 pp; clothbound ISBN 0–8412–0906–5; paperback ISBN 0–8412–0933–2

Developing a Chemical Hygiene Plan
By Jay A. Young, Warren K. Kingsley, and George H. Wahl, Jr.
paperback ISBN 0–8412–1876–5

Introduction to Microwave Sample Preparation: Theory and Practice
Edited by H. M. Kingston and Lois B. Jassie
263 pp; clothbound ISBN 0–8412–1450–6

Principles of Environmental Sampling
Edited by Lawrence H. Keith
ACS Professional Reference Book; 458 pp;
clothbound ISBN 0–8412–1173–6; paperback ISBN 0–8412–1437–9

Biotechnology and Materials Science: Chemistry for the Future
Edited by Mary L. Good (Jacqueline K. Barton, Associate Editor)
135 pp; clothbound ISBN 0–8412–1472–7; paperback ISBN 0–8412–1473–5

For further information and a free catalog of ACS books, contact:
American Chemical Society
Distribution Office, Department 225
1155 16th Street, NW, Washington, DC 20036
Telephone 800–227–5558